THE X IN SEX

DAVID BAINBRIDGE

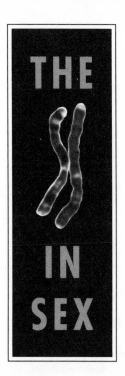

THE

IN

SEX

HOW THE X CHROMOSOME
CONTROLS OUR LIVES

HARVARD UNIVERSITY PRESS
CAMBRIDGE, MASSACHUSETTS
LONDON, ENGLAND
2003

First Harvard University Press paperback edition, 2004

Library of Congress Cataloging-in-Publication Data
Bainbridge, David.
The X in sex : how the X chromosome controls our lives / David Bainbridge.
p. cm.
Includes bibliographical references and index.
ISBN 0-674-01028-0 (cloth : alk. paper)
ISBN 0-674-01621-1 (pbk.)
1. X chromosome. 2. Sex determination, Genetic. I. Title.

QH600.5 .B35 2003
611'.01816—dc21   2002027367

For Edward

Thanks for taking the Y off my hands.
I wasn't really using it much anyway.

And if you have any complaints about the X, blame your mother.

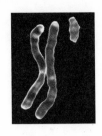

CONTENTS

A woman is, as it were, a mutilated man, for it is through a certain incapacity that the female is female.

Aristotle, *Generation of Animals*

Thus man . . . is the active principle while woman is the passive principle because she remains undeveloped in her unity.

Hegel, *Philosophy of Nature*

One is not born, but rather becomes a woman.

de Beauvoir, *The Second Sex*

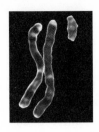

PROLOGUE

It has been six weeks now. Six weeks of tireless, frenzied activity since that sperm jostled its way into that egg. So little time spent in this warm, dark, womby home, and so much achieved.

First came those days of dividing—ceaseless splitting and splitting that turned a single fertilized egg into hundreds of tiny cells. Most of those cells went into making a protective hollow ball, but safe inside that ball lay another clump of cells with a very special destiny. That was four weeks ago, but how things have changed. The little clump of cells now looks recognizably like a tiny baby. It has grown, of course, but even more dramatic is the way it has organized itself into a child, with head, body, arms, legs, eyes, mouth. Really, all these bits and pieces have to do in the remaining thirty-four weeks before birth, is grow.

Yet there is one part of the baby that is still far from finished, and in many ways it is the most important. After all, the reason why nature makes us have babies is so that those babies can have their own babies. But even though so much of the baby's body has been mapped out, the parts that will make future grandchildren are far from complete. Deep inside the embryo's belly, just to the side of its kidneys, lie its gonads—the organs that will eventually turn into ovaries or testicles and drive the child's sexuality up to birth, to puberty, and long into adulthood. The gonads are already crammed full of germ cells—

future sperm or eggs—but at six weeks, they still do not know which way to turn. The embryo does not yet know if it is a boy or a girl.

The jargon name for the gonads at this stage is "indifferent," and somehow it describes them so well. They are, in fact, supremely indifferent. They are neither male nor female, but hovering in sexual limbo somewhere in between. All the rest of the sexual organs are equally undecided; simple tubular structures awaiting their cue to turn into male or female genitals.

So what is the spark of sexuality that makes a child a boy or a girl? In this book, I hope I can show you how this spark drives us to become men or women—people apparently so different, but made from the same stuff. Although many people are aware of the principles by which we are allocated our sex, I suspect that few realize that this is a story rich in history, evolution, and philosophy that challenges our views of society. We humans use an unusual method to decide our gender, and it can have dramatic effects on the way we live our lives. It may help many of us become "normal" men and women, but it also consigns many to a life of disease, disrupts the everyday running of our body, and even forces women to live a bizarrely double life.

The actual physical entity that causes all this upheaval is a little nugget of life called the X chromosome, and this is its story.

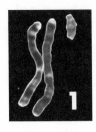

MAKING A DIFFERENCE

What did it all mean? Hermann Henking peered intently down the microscope and desperately tried to explain the strange behavior of the little purple flecks on the glass slide. It was 1890, and all over the world scientists were studying these flecks, yet still they seemed as mysterious as ever. Many believed that they held the key to the holy grail of biology—the mechanism that allows offspring to inherit characteristics from their parents—but no one could prove it.

As Henking sat in his dingy laboratory in Leipzig, he was about to make a discovery that was to revolutionize the science of inheritance, but he was also to set that science back ten years. His observations would lead directly to our modern understanding of sexuality and inheritance, but no one would realize the importance of his work for another decade. Indeed, Henking himself was to wait a further thirty years before another zoologist explained to him the magnitude of what he had achieved.

The 1890s were, in retrospect, a deeply frustrating decade. Charles Darwin had already shown, in his book *On The Origin of Species,* that external forces could make animal species change over time, so long as parents could bequeath characteristics to their offspring. Yet no one could say how all this bequeathing was done. Everyone knew that children often resembled their parents, and that the same seemed to be true of calves, foals, puppies, kittens—in fact, if you hunted hard enough, family resemblance seemed to occur in any type of animal you cared to look at.

So inheritance seemed to be a ubiquitous phenomenon, and yet nobody could explain it. Many zoologists argued about which bits of sperm and eggs carried the instructions to make each generation resemble the last. Most agreed that there must be some physical means by which these instructions were imparted to a new individual, but a few still argued that inheritance was a spiritual, and not a physical process—to believe that inheritance was somehow mundane might remove God from his place at the center of human affairs. And here was Henking, making a discovery that was to resolve this whole debate. He was finding the inheritance machine, and he did not even realize it.

Although it may sound rather unpromising, Henking's strained eye was fixed on chromosomes from the testicle of an insect called *Pyrrhocoris*. Twenty-five years earlier, a new microscopic dye had brought the chromosomes ("colored bodies") into view for a generation of scientists hunting the inheritance machine. These chromosomes had been found to behave in an enticingly unusual way. Every cell in the body holds a set of chromosomes, and most of the time these chromosomes do not, to be honest, look very special. The time when chromosomes spring to life is when a cell starts to divide into two daughter cells. The chromosomes magically change from rather vague, diffuse things into a neat set of clearly defined little threads carefully arranged in the middle of the cell. Even more intriguingly, the chromosomes are then teased apart into two apparently identical bundles, one of which is then packaged into each of the daughter cells. It seemed that chromosomes must be doing something pretty important if they were so neatly partitioned between cells, so could they be the inheritance instructions that scientists sought?

Henking was not studying fly testicles because he was interested in fly testicles. Scientists can be strange people, but not usually that strange. One of the most intriguing features of chromosomes was apparent only in testicles and ovaries, where cells divide to make sperm or egg cells—the cells that make the next generation. The cell divisions that made sperm and eggs seemed very different from the divisions that made all other cells. Instead of the single-step cell division that takes place when other cells are made, the chromosomes go through a strictly choreographed sequence of two successive divisions, and during these divisions, they are in a state of constant interaction with each other. Why should there be such an especially elaborate chromosomal dance when eggs and sperm are made, if chromosomes are not involved in inheritance?

It was at this point that Henking made the observation that was to etch his name forever in the history of biology. He noticed that during the two-part chromosomal dance in the testicle of *Pyrrhocoris,* there was one chromosome that stayed slightly to one side and did not take part in the festivities. As a result of its reticence, at the end of the dance it was not divided equally between the progeny cells—it only ended up in half of the sperm. It looked just like all the other chromosomes, but Henking could not understand why it did not join in. Mystified by this "wallflower chromosome," he gave it a name, although we are not entirely sure why he chose that name. Perhaps because it was mysterious, or because it appeared to be "extra," or because it was an apparently redundant "ex-chromosome," he called it "X."

The name stuck. To this day, most other chromosomes get simple, boring numbers for names, but X remains enigmatic X. It *was* special, but much more special than Henking had realized.

Essence of Man, Essence of Woman

Male and female. He and she. Impregnator and gestator. The duality of the human race is such an inescapable part of our existence that throughout much of history few have even wondered why it is so. Just as every person is irrevocably allocated a sex at birth, so is sexuality itself woven through the whole of human life.

In every human society men and women have had different roles, different destinies, and the same was true of the animals they herded or hunted. One thing is especially clear: males and females are more different than they really need to be simply to play their different roles in reproduction. Why do women look so different from men, think so differently to men? It was obvious that sexuality is not just restricted to the loins—it permeates the whole body. A young woman's hand is not like a young man's hand, but why, and how? Why should the sexes be so deliciously, unnecessarily different?

Although the reasons for the differences between the sexes were rarely explicitly questioned, they pervaded mythologies throughout the ancient world. For every Mars there was a Venus, for every Eve an Adam. The father sky–mother earth myth appeared around the world with bewildering frequency. Although female and male were not always assigned equal rank, the duality of sex was ingrained into almost every polytheistic tradition. Good relations between the sexes were obviously essential if people were not to become barren, or even starve to death. Sexuality was a necessary obsession in the ancient world, and shining like a beacon of reason among all this mythological fervor is a single figure at the source of our current understanding of the sexes: Aristotle.

Reading Aristotle's *Generation of Animals* is to come face to face with a man in the process of building a new way of thinking about the world. His seems a very modern mind, taking anecdote and personal observation and trying to explain the cause of what he saw. It was this obsession with cause that led him to crystallize the central question of sexual biology: why and how are man and woman made different?

For such an antique chap, Aristotle's thinking and writing seem incredibly liberal, even by modern standards. He was free to discuss the relative roles of men and women, to describe the intimacies of human sex, and even to consider humans as just another animal. Ancient Greece must have been a liberating place. Yet there was one issue that even Aristotle could not resolve. It may have seemed rather esoteric at the time, and it may seem so to you now, but as I will explain it remains the central philosophical question in our twenty-first-century debates about sexual biology and the roles of the sexes. At one point in the *Generation of Animals,* Aristotle states that "the female is opposite to the male," as if the sexes are diametrically opposed counterparts that together make up the human species. Yet elsewhere he describes women as a somehow modified form of men, as in the rather politically incorrect quotation with which I began this book.

So are the sexes truly opposite—essentially different in nature? Or are they more subtle alterations to a common mold? This is not just playing with words, as these questions not only inform our investigations of the mechanisms of sex, but they also color the way that the sexes treat each other. In the Judeo-Christian Bible, for example, it is made clear that Adam was the original human, and that Eve was generated from a ribby offcut, rather as an afterthought. The implication was clear: women are a derived, sullied form of the pure masculine source. Of course this is just a myth, but for the last two thousand years most Europeans have believed it.

Today it seems that Aristotle's indecision over whether men and women are truly opposites or just complementary in some way was wise. He was in no position to resolve the issue and he probably knew it. Instead, he paved the way for the intellectual study of the sexes—he dedicated a large portion of the *Generation of Animals* to the question of why some babies become boys and some become girls. Perhaps he hoped that by resolving the mechanism of what we now call sex determination, he would edge closer to understanding the nature of male and female. Whatever his reasons, his ideas, almost without exception, still have striking resonance in modern biology.

Aristotle's first observation was that embryos do not seem to actually have a sex until surprisingly late in their development (about six weeks in humans). To Aristotle, this raised the question of whether an embryo is in any sense sexed in its first few weeks of life. He argued that the apparent sexual indecision of early embryos meant that they may not be assigned a sex at the moment of conception, but that it might somehow develop later. Alternatively, there might be some "seed" of maleness or femaleness that was present from conception, but which simply was not evident to the naked eye. The time at which a baby was assigned its sex lay at the heart of understanding sexual differentiation, and it was to be fought over until the early twentieth century.

One idea, championed by Anaxagoras, stated that the baby's eventual sex was decided at conception—some sort of germ of sexuality was indeed instilled at the very moment of the child's creation. Because Anaxagoras believed that the seed of the baby came from the father and that the mother simply provided a place for it to grow, he asserted that the sex of the baby was determined, in advance, entirely by the father. Wrong reason, right answer. As I will explain, a child's sex is indeed controlled by a factor arriving in the paternal "seed."

This idea of predetermination of a child's sex went even further. It was claimed that men lacking one testicle could only sire children of a certain sex, and so male-making seeds were said to come from the father's right testicle and female-making seeds from the left. Of course, this may now seem a strange suggestion to us, but it introduced an important idea: that men make two different kinds of sperm to make male and female children.

The opposite theory, proposed by Empedocles, was that an embryo is entirely sexless at conception, but that it becomes a boy or a girl because of its environment—he claimed that a hot womb would yield a boy and a cold womb a girl. There were also many other external influences that were claimed to influence the sex of children, none of which seem to amount to more than folklore: the weather, or the temperature of drinking water, or the timing of copulation. Although Aristotle did not think much of the idea that a child's sex was decided by its environment—for a start, it could not explain how women can conceive twins of different sexes—the idea was to become incredibly influential, and indeed it was to hold sway until the mechanics of sexual differentiation were finally discovered. Yet here again I will argue later that the idea that environment can control our sex contains a germ of truth.

Aristotle himself opted for a third way by which sex might be instilled into a child. Propounded by Democritus, it involved a struggle between the generative forces of father and mother. The idea was that male semen tries to make a child male, whereas female semen (whatever that is) tries to make a child female. Thus, a child's sex is decided on the basis of which parent's semen prevails in a bizarre interseminal competition. It is, indeed, a very democratic idea in which a crucial outcome is decided by a fair contest.

There the matter rested for over two thousand years: competing claims about how our sex is decided, all equally lacking in proof. Father, mother, environment—which would turn out to be the trigger to babies' sex? Despite the lack of concrete answers, biology was hardly static during all this time. The seventeenth century brought the invention of the microscope and the revelation that the male "seed" was, in fact, myriad little tadpole-like "seed-animals" or "spermatozoa." Not until the nineteenth century was the elusive female semen discovered, the precious human eggs lying hidden in the ovaries. Wonderful things, to be sure, but still yielding no clues to help find the motor of sex.

All this time, an awful worry clawed at the minds of biologists, just as it had plagued Aristotle: how can a child inherit its sex? You can inherit blond hair from your parents, especially if they are fair, but why do some babies inherit maleness from their parents and others femaleness? We all have a woman and a man for our parents, so what is there in our heritage that pushes us one way or the other? Many thought it unlikely that our very femininity or masculinity could be inherited in the same way as other traits. Yet one thing seemed clear: science was unlikely to unravel inheritance without first unraveling sex.

And Here's the Reason Y

By the beginning of the twentieth century, ten years after Henking first spied his antisocial X chromosome, chromosomes were very much in vogue. Many people were coming to believe that they were the physical mechanism by which traits were inherited: springing into view just before cells divide, and neatly segregating between daughter cells, they looked like perfect candidates for the magical inheritance machine. But few at the time can have predicted how the story of chromosomes was about to become intertwined with Aristotle's confusion over the inheritance of sex.

Around this time, rumors appeared of a remarkable discovery published by an obscure monk over thirty years previously. Word was out that this discovery could change our understanding of inheritance forever. Working with pea plants, the Silesian friar Gregor Mendel had shown that two parent plants pass on certain traits, such as color or height, in neat mathematical ratios—50:50 red:white offspring, or 75:25 tall:short offspring. He had not discovered the machinery of inheritance, but in his own quiet way he had shown its end results— the pattern of inheritance. Mendel's work was a tremendous fillip for everyone who believed that inheritance was a mundane physical process—a simple random allocation of traits. Or rather, it would have been, had he not published it in the delightfully obscure *Journal of the Brno Natural History Society,* which probably explains why it was ignored from its publication in 1866 until the beginning of the next century. Scientists had desperately sought this kind of evidence for decades, and one can imagine their reaction when they found that it had been gathering dust on central European bookshelves all along.

With such clear evidence of a physical process of inheritance, as well as those chromosomes dutifully dancing at every cell division during sperm and egg formation, the race was on to link the two. The proposition of such an un-Godly mechanism for the dissemination of characteristics from parents to children horrified the religious, but the physical inheritance bandwagon was now unstoppable. Could the unspeakable really be true? Did animals and, heaven forbid, people just have little machines in their cells that distributed their beauty, their character, and even a little of their soul to their babies? Along with Darwin's irreligious but strangely compelling ideas about evolution, the concept of physical inheritance seemed to strike at the heart of Christian belief. The ancient Greeks, who suggested that a child might be produced entirely by a union of its father's semen and its mother's menstrual blood, at least had the excuse of being heathen. Now supposedly Christian people were saying that God played no part in the most miraculous part of human life—the generation of a new baby. Was nothing sacred?

Like many others, Clarence McClung believed that the strange behavior of the chromosomes belied a role in inheritance, but he was the first to suggest that Henking's X chromosome might actually have something to do with sex. Many people still called it the "accessory" chromosome, as if it was somehow superfluous and unnecessary, but McClung had big plans for it. From his own work on grasshoppers at the University of Kansas, he decided that Henking's X was not just some random idiosyncrasy of the bug he had been studying, but something much more important. In 1901, he published his ideas, claiming that the X chromosome was an excellent candidate for the sex inheritance machine because it is packaged into some sperm but not others. McClung argued that the arrival of an X-bearing sperm at the egg led to male offspring and that an X-lacking sperm would generate a female. Thus he firmly identified the X as being "male determining."

This confident assertion of the physical, chromosomal inheritance of sex caused a furious argument in the first decade of the twentieth century, and it was not just religious conservatives who did not like McClung's ideas. Many scientists saw his theory as being far from proved. He had not actually shown that the X conveys maleness, nor had he shown that the converse was untrue: that the X chromosome is some strange by-product of being male, and does not itself actually cause maleness. For a few short years, McClung's detractors could still point to their theories as being just as valid as McClung's vision of the X as the single switch to maleness. Many still supported the democratic idea that male and female factors somehow slug it out for control of the embryo's sexual destiny. Other scientists dug in with Empedocles, still touting environment as the controlling factor in sex. Even McClung himself slightly confused matters when he suggested that, although the sperm carries the sex-determining factor to the egg, the egg decides whether to let in X sperm or non-X sperm. He proposed that the egg selected different sperm on the basis of the environment, presumably to generate a child of the right sex to best serve the future needs of the species.

All this argument reached a sort of resolution in 1905 with a series of discoveries by Nettie Stevens at Bryn Mawr College for Women in Pennsylvania. As a man discovered the X chromosome, then perhaps it is rather appropriate that a woman discovered its counterpart, the Y. Born in Vermont eight weeks after the surrender of Fort Sumter in 1861, Stevens's career as a scientist was a story of immense tenacity in a male-dominated world. Working as a librarian and a teacher to raise the money to go to Stanford University, Stevens was forty-two when she was finally awarded her doctorate at Bryn Mawr. Between 1903 and her death from breast cancer in 1912, she hurled her intellectual energy at the question of chromosomal inheritance in insects—in her case the lowly mealworm *Tenebrio*.

The mealworm had an important story to tell, but only to someone as dedicated as Stevens, who was able to count every chromosome in its cells. Count after count confirmed her suspicions: female mealworms' cells always had twenty full-size chromosomes, whereas males had nineteen large chromosomes, and one tiny little one. What's more, females produced eggs that always contained ten large chromosomes, but males' sperm could have either ten large ones, or nine large ones and a small one. She had discovered the great prize of heredity—the chromosomal basis of sex.

Central to this discovery was her idea that chromosomes exist in pairs, and that the tiny chromosome she had seen was, in fact, the partner of the larger X chromosome. This diminutive partner is now called the Y chromosome, and Nettie Stevens was the first person to realize that it is this, and not the X that controls sex. McClung had been on the right track, but he had got everything the wrong way round. The X chromosome had been discovered first because it was bigger, but it is its smaller counterpart that dictates the sex of meal-worms, many other insects, and humans as well. McClung had claimed that the X was male determining, but Stevens showed that this honor was reserved for the Y. Men and male mealworms are XY, and the Y is what makes them male, and women and female meal-worms are XX. Her results could also explain Henking's original dis-covery of the X chromosome standing aside from the chromosomal *melée* during sperm production: the single X chromosome in male cells has no equal partner with which to dance, so it sits to one side, waiting for the other chromosomes to finish the party.

Anaxagoras had been right all along. The sex of a child is indeed determined at the moment of conception by whether its father con-tributes an X sperm or a Y sperm to its creation. Mothers have no influence upon the sex of their children, although admittedly not for the almost chauvinistic reasons that Anaxogoras had suggested. Stevens was quick to realize that the process she had discovered could explain why boys and girls are born in roughly equal numbers—by equal packaging of X's and Y's into sperm inside the testicle. So sex was inherited after all, but in a special random way, with little control by either parent. The acquisition of our most important trait had turned out to resemble the tossing of a coin.

It all seemed too good to be true—two crucial problems solved at a stroke. Not only had Stevens's work shown how sex was inherited, but it was also the first demonstration of any trait being bequeathed to future generations via chromosomes. By solving the problem of sex, she had also gone a long way toward demonstrating that chromo-somes are the inheritance machine. Indeed, within five years, the sex chromosomes were to produce irrefutable proof that other, far less dramatic traits are also transmitted from generation to generation on chromosomes. Experiments with the fruit fly *Drosophila* showed that the pattern of inheritance of eye color can only be explained if it too is inherited on a sex chromosome, in this case the X. So early twentieth-century science had effectively answered Aristotle's two biggest questions simultaneously—how do children inherit characteristics, and how are they made male and female?

Yet the discovery of the macho Y chromosome had rather eclipsed the heroine of our story, the X chromosome. XX women and XY men were all very well, but where did they leave the X chromosome? Y, tiny as it is, is the very essence of masculine—if you have it then you should be a man—but where does the essence of feminine reside? X chromosomes do not seem to make women; it is the lack of a Y that makes a baby girl. Is the X chromosome just there to pad out a cell's chromosomes when there is no Y around, or does it have a more positive role?

Do not worry about the X's status—the relative sizes of the two chromosomes should hearten us. The X is a bold, full size, *bona fide* chromosome, whereas the Y is a sad, shrunken, vestigial thing. The X chromosome may not be the crude arbiter of male and female, but it has hidden depths, and a very special place in controlling our lives, both for men and for women. In the rest of this book I hope to convince you that it is just about the most compelling little scrap of stuff in existence.

The Great Chain of Being . . . Sexy

It is all very well to say that two X chromosomes make you a woman and that an X and a Y make you a man, but why should this be so? Most people happily accept this XX/XY story, but it is a great leap from having a couple of little blobs in your cells to having a male or female appearance and mind. Our sexuality pervades our entire body, so how do these tiny chromosomes make their presence felt throughout us?

The answer to this question lies with that little six-week-old embryo that we visited in the Prologue, floating in its tiny protective fluid sac. For all the world it looks sexless, and indeed babies that will eventually be boys and girls are indistinguishable at this early stage of development. Even its gonads give us no clue—not yet testicles or ovaries, they are still undecided, "indifferent." So what is it, exactly, that will spur this little life to be one sex or the other? What forces will act on it to cast it into a woman, who can bear her own children, or into a man, who can decide the sex of his?

Since Nettie Stevens's discovery, more evidence has come to light that it is indeed not the X, but the Y chromosome that decides the sex of human children. Much of this evidence has come from studying children born with an abnormal number of sex chromosomes. Sex chromosome abnormalities are surprisingly common, and I will return to them later in this book. For now, suffice it to say that inheriting an abnormal number of sex chromosomes is often not incompatible with life, and many carriers survive into adulthood. Some may have a rather characteristic appearance, some have learning difficulties, and many are infertile, although many never even realize that nature has overendowed them with its seeds of sexiness.

For a start, sex chromosome abnormalities tell us that men are not men because they have only one X. Occasionally babies are born who have only one X chromosome and no Y. These XO babies invariably look female, although after puberty they start to look different from XX women. A second important thing that these sex chromosome abnormalities tell us is that having a Y chromosome does indeed nearly always make you male. Whatever else is going on in the jumble of sex chromosomes, even in XXY, XXXY, XXXXY, XXYY, or XYY babies, the presence of at least one Y seems to be sufficient to make an obviously male child. No matter how many extra X's there are to force the issue, a single Y can usually win through and make a child into a boy.

The Y, it seems, is king. These are difficult times for those of us who champion the X chromosome, but fear not—it will have its day.

Yet the apparent primacy of the Y chromosome still takes us no nearer to explaining how the sex chromosomes actually go about deciding the sex of a child. An important link in the chain connecting chromosomes to sexuality was discovered in the 1940s when Alfred Jost showed that the key organ controlling the embryo's decision about its sex is the testicle. Zoologists had known for some time that testicles and ovaries develop from the same sexless indifferent embryonic gonads, but Jost was the first to show just how many of the differences between males and females are driven by the developing testicle. In a series of experiments in rabbits, he showed that removal of the testicles from a male embryo led to the birth of a rabbit kitten with essentially female anatomy—externally and internally. The sole exception to this conversion was that the kitten lacked ovaries. Also, testicles transplanted into female embryos drive them to develop male anatomy. Testicles make boys into boys, and girls become girls because they do not have testicles.

In a single coup Jost had split the question of developing sexuality in two. Where once scientists had wondered how the sex chromosomes turn babies into boys or girls, they now had two sequential questions to answer. First, how does the Y chromosome make testicles and second, how do testicles divert embryos from being girls into being boys? The seemingly intractable chain—the question that dated back to Aristotle—was at last being broken into manageable bits.

The first bit of the chain—how a Y chromosome makes a testicle—was to wait until the early 1990s for a more complete explanation. Once again, in the face of such a difficult problem, reproductive biologists had to return to nature's own laboratory for some hints. They found that, very occasionally, in perhaps 0.005 percent of babies, the rules of sex are apparently flouted, leading to the birth of XX boys or

XY girls. This sex reversal is a very rare event—so rare, in fact, that it is really quite surprising that it was ever discovered at all. Of course, in the great scheme of things, these babies are numerically insignificant, but to aficionados of the embryonic sex decision, they are sheer gold dust. It is their very rarity that makes sex-reversed people so fascinating. The Y-makes-boy rule is an enticing one because it is almost always followed. Yet it was its occasional failure that made scientists sense a weakness in the rule—a weakness that might help them understand how it worked.

Researchers meticulously picked apart the X and Y chromosomes of sex-reversed people. Soon it became clear that those apparently normal, albeit misplaced, sex chromosomes were not all that they had seemed to be. They were, quite simply, damaged. The Y chromosome in XY girls seemed to have a piece missing, and one of the X chromosomes from those unexpected XX boys had an extra fragment tacked on. Most remarkably of all, the missing piece in the XY girls seemed to be suspiciously similar to the extra piece in the XX boys. This tiny fragment, normally present on the Y chromosome, appeared to be sufficient in itself to make a child male. Without it, the Y does not work, and the baby becomes a girl. With it, the X chromosome suddenly drives XX embryos to become boys.

By studying more and more XY women and XX men, and comparing the fragments that they respectively lacked or possessed, scientists gradually hunted down the male-causing region of the Y chromosome. After some premature celebrations when an impostor named *Zfy* was briefly hailed as the trigger to maleness, the true keeper of the keys to masculinity was discovered, a gene called *Sry*, an abbreviation of the rather unpoetic "sex-determining region on the Y chromosome."

Did I manage to slip in the word "gene" unnoticed? It is a word rich in meaning and heavy in connotation—something rather rare for the lackluster world of scientific terminology—and I do not want to do it an injustice here. Yet I also do not want to bog down our sex-laden story in the minutiae of the discovery, mechanics, and chemistry of genes. I will return to genes after this chapter is finished, but for now you can get by with a fairly skimpy understanding of what they are. In short, each gene is an instruction manual that a cell can use to make something useful. Usually that useful thing is a protein. Obviously, you need lots of useful things to make a baby, so we need lots of genes—tens of thousands in fact—and these must be neatly stored somewhere. Our genes are stored, as you may have guessed by now, in our chromosomes, and that is why we inherit the instructions to make our body on our chromosomes. So a gene is an instruction manual. That is enough for now.

So in discovering the *Sry* gene, scientists had done something rather wonderful. They had found a little scrap of chromosome capable of controlling the fate of an entire baby. *Sry* was a single gene that seemed to control the development of a whole organ system, and eventually a whole child—no one had ever discovered anything nearly as special as that before, and they probably cannot be said to have discovered anything similar since.

But is *Sry* as amazingly influential as it seemed? Within a few years, with a little genetic jiggery-pokery in mice, researchers were soon to show that it probably is. First of all, they looked to see what *Sry* is actually doing in that early "indifferent" gonad. In female mouse embryos, they found no evidence that the *Sry* instruction manual was being used at all. This is hardly surprising. Females, being XX, do not have an *Sry* gene. When they looked in the gonads of early male mouse embryos, however, there was a flutter of protein production from the *Sry* gene immediately before the gonad became recognizably testicle-like. The *Sry* protein is only made in the gonad, and only for a couple of days, and then it fades away, but that is enough. That little flutter of activity is sufficient to drive a cascade of changes that will alter the embryo's future forever, and pretty much the same thing is now known to take place in human baby boys as well.

Next came the final confirmation of the macho power of *Sry*. Scientists were soon able to show that if the gene is deliberately damaged in XY mouse embryos, then they are born with female anatomy. Even more remarkably, they were also able to demonstrate the opposite: if *Sry* is added onto the chromosomes of XX embryos, then they are born looking male—a discovery that led to a picture of a friendly-looking XX male mouse proudly displaying his genitals to the waiting world on the cover of the esteemed scientific journal *Nature*. What the scientists were creating, in fact, was the mousy version of sex-reversed people. They had re-created nature's transgender experiment and, in the process, had shown that this single gene is also the key to human, as well as mouse, sexuality.

So now everyone knew that *Sry* is the switch that is flicked to make a boy. But scientists are a tenacious bunch and once they have their gene, they never let it go. Just having the switch that turns on the testicle is not enough. Since the discovery of *Sry,* they have researched how, when, and where *Sry* actually acts to make that testicle. In effect, they have been trying to work their way along the rest of the first half of Jost's bisected chain, between the Y chromosome and the testicle. They sought more genes and they found more genes. Nothing in biology is ever as simple as one would like, and the search has turned up an intimidating array of other genes that are switched on by, or interact with, *Sry*—an initially meaningless list including such enigmas as *Dax1, Gata4, MIS, Sf1, Sox9, Wnt1,* and *Wt1*.

Just when scientists thought they had the whole thing sorted out, this jumble of new genes appears. If *Sry* is the gene that does the job, then what are all the other ones for? Well, for a start, some of them really do seem to form a neat little chain connecting *Sry* to the testicle. The *Sry* gene seems to make a protein that switches on the *Sox9* gene, and *Sox9* then makes a protein that switches on *MIS*. *MIS* is especially important, for it is believed to coordinate the construction of the testicle itself. In fact, we almost seem to have our little half-chain mapped out:

Y chromosome → *Sry* → *Sox9* → *MIS* → testicle

Of course, life is never that simple. First of all, there are several genes on the list that do not slot simply into this little plan, as well as lots of others that geneticists have probably not even discovered yet. All these other genes are now thought to control the way that the chain of events works, probably in a bewilderingly complex way. Another problem is that most of these genes seem to do lots of things other than making boys into boys. Whereas a child that lacks *Sry* simply turns into a healthy girl, a child that lacks *Sox9* is born severely deformed. In other words, making boys is a full-time job for *Sry*, but it is only a part-time job for *Sox9*—it spends most of its time doing other things. Not only do *Sox9* and lots of the other genes have other, less sexy jobs, but many of these genes are not even carried on the Y chromosome either. Most of them sit on other chromosomes, waiting for *Sry* to trigger them from afar.

There I must leave the first part of the chain, still slightly tangled. All we can say for sure is that boys grow testicles because their Y chromosome carries an *Sry* gene that, one way or another, makes their gonads turn into testicles, the powerhouses of boyhood. But now comes the question of just why those testicles are the powerhouses of boyhood. We must piece together the second half of Jost's broken chain:

testicle → ? → ? → little boy

Just as nothing is ever simple in biology, I would claim that nothing is ever worked out in the right order, either. Whereas the first half of the chain was explained in the late twentieth century, teasing information from the second half of the chain began nearly a half-century ago. Working out how testicles make boys was apparently easier than working out how Y chromosomes make testicles, especially before the advent of modern genetic trickery.

Testicles make boys by releasing hormones, and scientists have been able to study hormones for a lot longer than they have known how to study genes—hormones are really rather more tangible. Instead of being strange instruction manuals on chromosomes, hormones are just chemicals made by one part of the body, which are then cast adrift in the blood to make landfall in some different, distant part of the body. Once ashore, the hormones then make these far-flung parts of the body do all sorts of unusual things. And the hormones made by the testicle carry an unequivocal message around the body: "make a boy."

Perhaps the first hormone to start to mold an XY embryo into a boy is the rather inelegantly titled Mullerian inhibiting substance (made by that *MIS* gene I mentioned before). Mullerian inhibiting substance is the next link in the chain after the testicles begin to form. Indeed, one of its most important roles is to help complete those testicles. Testicles are complex things, full of hormone-producing cells and sperm-nurturing tubes, and they have to be very nearly finished by birth if the boy is to be fertile when he grows up. Indeed, scientists now think that the level of a man's fertility is pretty much set by the time he is born—this is one of the reasons why people worry about the risks of female hormone–like pollutants and what they might be doing to male babies' testicles while they are still inside the womb.

Another crucial job done by Mullerian inhibiting substance is the formation of the necessary tubes to make a boy. Boys need a thin tube to carry sperm from the testicles to the penis (the vas deferens), but girls need much wider-bore tubing to make their reproductive organs (Fallopian tubes, uterus, and vagina). The trick that the embryo pulled off with its gonads—using the same gonads to make either testicles or ovaries as required—does not seem to work with the tubes. Instead, while they are still "indifferent," embryos actually have two sets of tubes. Although it may seem a bit disconcerting, before six weeks of development we all have the forerunners of sperm tubes, Fallopian tubes, and a uterus. If anything, we are not sexless, but double-sexed. To become a boy or a girl, all we have to do is discard the set of tubes we do not require. The tubes that boys need are called the "Wolffian" ducts and the tubes that girls need are called the "Mullerian" ducts. This is how Mullerian inhibiting substance got its name—one of its main jobs in boys is to destroy the Mullerian ducts. Men do not need a uterus and this hormone is what rids them of it.

At the risk of stating the obvious, another big difference between women and men is that they keep their gonads in different places. Women, of course, discreetly closet their ovaries away in their abdomen, whereas men proudly dangle them in the fresh air. No one knows why men have such an exhibitionist approach to gonad deportment, but it does seem to be essential for their fertility. If a testicle is accidentally imprisoned inside the abdomen during a boy's development, it is usually kept too warm to produce healthy sperm—although this does not explain why whales, dolphins, and elephants can sire future generations perfectly well with their testicles tucked in their balmy abdomens.

Anyway, in humans at least, gonads must end up in the right place, and the testicle manufactures a hormone to make sure this happens. This hormone is called insulin-like-hormone-3, and it works its magic by controlling the tethers that hold the gonad in place. At first the gonad is tied next to the kidney by a short strap called the suspensory ligament. There is another longer, looser tether that attaches to the gonad, and this has a name to conjure with: the gubernaculum. Insulin-like-hormone-3 strengthens the gubernaculum in male embryos so that it starts to pull the developing testicle away from the kidney, and the testicle's enforced migration does not usually finish until it has popped out of the abdomen altogether, and landed in the scrotum. In other words, the testicle makes the hormone that guides it to its eventual well-ventilated destination.

For this descent into the scrotum to occur, the other tether, the suspensory ligament, has to be weakened, and the testicle makes a final type of hormone to do just this. Androgens are a group of hormones of which by far the most well known is testosterone, and I will soon explain why androgens are so famous. But for now they start fairly timidly, by weakening the suspensory ligament.

Soon, however, the androgens really get to work. It is their influence which makes the male Wolffian ducts grow into the tubes that carry sperm from testicles to penis. Also, and most dramatically, they make baby boys become externally male. Before androgens kick in, the embryo's external genitals look just like a girl's. But under the influence of androgens, everything soon gets rearranged. First, the lips around the bladder outlet fuse and become saggy bags into which the testicles fall—the scrotum. Then, the bladder needs a new way to drain out into the outside world, so a new tunnel is dug to the tip of the penis, which is itself growing under the command of androgens.

The reason that androgens, and testosterone especially, are a household name—whereas Mullerian inhibiting substance and insulin-like-hormone-3 are certainly not—is that they drive maleness after birth and well into adulthood. This is why testosterone carries so many connotations of virility. If you want to know what testosterone and its kin actually do, then just think of some of the differences between baby boys and adult men. There is still a lot of sexualizing to be done once a boy is born—after all, baby boys are not born with hairy chests and deep voices. Instead, these attributes have to be lovingly ingrained into them by androgens. In many animals, testosterone is also in control of males' brains, driving aggression and libido. Humans are a little more subtle than that, however, and although male hormones are important in giving men characteristically male brains, with small, but very real anatomical differences from women's brains, by the time they are mature, an adult brain's maleness can work pretty much autonomously, and not at the whim of androgens.

So there you have it: a boy. Perhaps you will have noticed that my description thus far seems a bit male-oriented—genes on the Y chromosome heroically wresting the child away from its passive female side to a bright, aggressive, male future. The march of maleness seems so purposeful and defiant—is human sexuality simply about deciding whether to be male or not? Is there only a choice between man and non-man? Is there no positive, active urge to become a woman? Well, what can I say? I am afraid that at first sight, this rather misogynistic scheme does seem to be how boys and girls are made different. Anaxagoras would have approved of such a scheme, in which the father's seed decides whether or not to make a precious son.

But first impressions can be misleading. Half of us are XX, half of us have ovaries, and half of us are girls. Is nature really likely to have consigned half of us to the humiliation of being an afterthought? Bear with me, and we'll see how the Y/testicle/male doctrine can be picked apart.

So Are Girls Just an Afterthought?

There is another way, of course, and for our unsexed embryo and its indeterminate gonads this other way is to turn into a girl. Many babies travel this route, yet their journey seems very different from that of boys. Almost all girls start their existence with two X chromosomes, instead of an X and a Y, and this is what seems to set them on their road to femininity. We saw how boys' chromosomes set off a chain reaction that makes them into boys, but is there an equivalent chain that makes babies into girls?

Perhaps the first sign that things are going to be very different from boys is that X chromosomes do not actually start the feminizing process at all. As I mentioned earlier, two X chromosomes do not necessarily make a girl. Evidence from people with abnormal numbers of sex chromosomes shows that being XXY makes people look male and that having just one X chromosome makes people look female, albeit slightly different from most women. So having two X chromosomes does not make a woman, even though almost all women do, in fact, have two X's. In fact, the only reason that having two X's often makes babies female is that it usually means that they do not have a Y chromosome, which is rather a negative reason.

Think of sex as a restaurant, with sex chromosomes for customers. This may not be the kind of restaurant you want to eat in, but bear with me. It seems to me that people often eat in restaurants in pairs, and indeed sex chromosomes normally travel in pairs as well. Now, imagine that X chromosomes are represented by nonsmokers and Y chromosomes by smokers. A couple enter the restaurant and the *maitre d'hôtel* asks them, "smoking or nonsmoking?" (For the sake of this analogy, I will assume that this is a European restaurant where the jackboot of political correctness has not yet crushed the spirit of the nicotine-addicted.) One of the couple smokes (XY), and so they have to go and sit in the smoking area (boy). Another couple comes in, and they say that neither of them smoke (XX). The *maitre d'* likes their faces and graciously allows them to take their seats in the nonsmoking area (girl). For the rest of the day, couples drift into the restaurant and are seated as appropriate by the draconian management, although strangely enough no couples appear who both smoke (YY)—perhaps these couples die of smoking-related diseases on the way to dinner.

Then, all of a sudden, some rather unusual people appear. First, a lone nonsmoker (XO) ambles in, and against his better judgment, the *maitre d'* cannot find any reason not to place that person in the nonsmoking area (girl). Then, an unruly group comprising three nonsmokers and a smoker (XXXY) burst through the door and they are hurriedly hidden away in the smoking area (boy), next to the toilets.

The reason I am dwelling on this regimented restaurant is that it demonstrates the indirect role of the X chromosome; the nonsmoker. Being a nonsmoker is not, in itself, enough to get you a place in the nonsmoking section, but in a world where most people dine as couples, finding a nonsmoking friend will get you the tobacco-free air you desire. In the same way, as long as an embryo plays by the rules and has two chromosomes, then if both of them are X, that embryo should become a girl. Being XX only makes you a woman in as far as it usually precludes you from having a Y. So unlike Y, the X chromosome has an indirect, conditional effect on a baby's sex.

In other words, at the risk of switching metaphors too quickly, we can see that the first part of our chain has turned out to be really rather vague. Having two X chromosomes usually means that a baby will become a girl, but apparently not because of any active intervention from either of those chromosomes. In the absence of any positive driving force, how then do girls turn their gonads into ovaries? For a start, if the X chromosome has no active girl-making function, then there is little point in looking for a female equivalent to *Sry*. There does not seem to be a simple, single switch to make an embryo into a girl.

One possibility is that the gonad simply turns into an ovary by default. The early indifferent gonad looks no more like an ovary than it looks like a testicle, so at first sight there may seem to be no particular reason why it should just passively drift into becoming an ovary. But perhaps we have been misled by our modern sense of gender

equality, or even by Democritus' old idea that a baby's sex is determined by some even-handed contest between male and female influences. In fact, various devious experiments with chopping and
changing the *Sry* gene in mice have shown that if *Sry* is not allowed to
act, or its appearance is significantly delayed, then the gonad indeed
simply assumes that it was meant to be an ovary all along. Lack of *Sry*
at the right time is all that is needed to start the embryo on the path to
being a woman. The gonad does, after all, become an ovary by default.
In the case of ovary development, delay and vacillation are what make
a girl.

Just because a gonad becomes an ovary by default does not mean
that the ovary is in any way a rudimentary testicle, even though this is
exactly how it was viewed until modern times. In his influential *Generation of Animals* of 1651, William Harvey, the man who discovered
the circulation of the blood, was probably articulating the beliefs of
many of his contemporaries when he said of the ovaries: "I, for my
part, greatly wonder how any one can believe that from parts so
imperfect and obscure, a fluid like the semen, so elaborate, concoct
and vivifying, can ever be produced, endowed with force and spirit
and generative influence adequate to overcome that of the male." Yet
the discovery of the microscope was soon to show that these "parts so
imperfect and obscure" embody an incredibly complex structure,
containing within it the eggs by which women contribute to the next
generation.

Scientists now know that eggs and sperm develop from the same
primordial germ cells in the indifferent gonad, and that it is the "sexualization" of that gonad that actually tells them what to become. Germ
cells in testicles remain small, and in adulthood they begin to divide
to produce the millions of sperm that a man can muster each day. The
fate of the equivalent cells in ovaries is very different. Long before a
girl is born she has created from her germ cells perhaps the most com-

plicated cells of all, human eggs. Constructed around these precious eggs is an array of cells dedicated to feeding and nurturing them, and when the time comes, these cells are also responsible for the waxing and waning of hormones that is such a feature of women's lives.

So the ovary is far from a failed testicle, even if it does form by default. In the absence of *Sry*, the gonads of XX embryos do not wait long. After a short pause, a very active process of ovarian construction begins. The tubules that would otherwise have gone into making the testicles are cleared away, and the eggs are made ready, along with their nurturing cells. The first major phase of making a girl is complete, but where was the chain of events that caused it to take place? Because of the lack of any specific girl-making trigger on the X chromosome, the formation of the ovary, crucial as it is, turns out to be a rather negative affair.

two X chromosomes → no Y → no *Sry* → construction of ovary

But now the ovary is in place, surely the development of the baby girl can take on a more purposeful air?

The gonad is starting to look recognizably like an ovary and the embryo has reached a turning point in its development. Now that it has ovaries and eggs, the embryo must form all its other girly paraphernalia if it is ever to be fertile. After all, there is no point in having ovaries without the correct plumbing. You might have thought that now would be the time for the ovary to take command, start pumping out some good strong hormones and to hurl the rest of the embryo to its female fate. And you would be wrong. Just as the X chromosome seemed inert at the time when the baby's sex was decided, the ovary seems similarly disinterested in the development of the rest of the baby girl. It almost makes you wonder if Harvey was right after all.

As I mentioned earlier, all embryos start off with two sets of tubes—the male Wolffian ducts and the female Mullerian ducts—and the testicle actively encourages the former and destroys the latter. The ovary is far less interventionist, and in fact all that can really be said of its contribution is that it does not behave like a testicle. It makes no Mullerian inhibiting substance so the Mullerian ducts stay where they are, and it makes no androgens so the Wolffian ducts fade away. So, again almost by default, the inertia of the ovary means that XX embryos are left with the tubes to make a vagina, uterus, and Fallopian tubes. Admittedly, later construction work takes place to complete these organs, but the fact that they are present at all is really the result of inaction, rather than action.

Next the ovary has to be kept tethered near the kidney, because ovaries are rarely found in a scrotum. Once again, it is lack of testicular hormones that ensures that the ovary ends up in the right place. It makes no androgens, so the suspensory ligament holding it inside the abdomen remains strong, and it makes no insulin-like-hormone-3, so the gubernaculum fades away before it has a chance to yank the ovary into the outside world. Once more, the ovary is remarkable not for what it does, but for what it does not do.

As with boys, the final part of the sexualizing process of girls is the formation of external genitalia, and even here the ovary takes a back seat. You may remember that I suggested that the genitals of an early embryo looked rather female. Indeed, they consist of a vulva-like opening, a recognizable pair of labia, and a small but clearly identifiable clitoris. Without the influence of androgens to forge them into a penis and scrotum, these genitals remain pretty much as they are. The ovary really does not have to do anything at all to make the embryo become externally female.

So the X chromosomes have all the opportunity in the world to take control of a girl's sexual destiny, but it seems that they singularly fail to do so. Some primordial sexual organs start off unsexed (the gonads), some are supplied in both male and female form (the tubes), and some come in a basic female model with an option to modify to a male variant (the external genitals). Yet whatever form these organs take, the X chromosomes never really seem to do anything to any of them. If the X chromosome is so apparently inactive, then what is it actually for? How can a so-called sex chromosome apparently play no part in controlling a baby's sex? Furthermore, if making boys is such an active process and making girls is such a passive one, how does this inform our views of men and women?

Whatever interpretation you place on the way boys and girls are made, one thing is for sure: the two sexes are not created by equivalent processes. Boys are the result of a neat chain of events reaching back to the inheritance of a Y chromosome, whereas girls are the result of that chain reaction not taking place. Is this really a true reflection of the X chromosome and femininity—a litany of negativity, delay, passivity, and inertia?

Ever since Jost's experiments began the half-century of research that has apparently confirmed Anaxagoras' and Harvey's chauvinism, this inequality of the generation of the sexes has fascinated doctors, zoologists, Marxists, religious fundamentalists, feminists, and even the occasional existentialist. Simone de Beauvoir mulled this idea over at length in *The Second Sex*. It certainly seems perverse that the latter half of the twentieth century, which brought unprecedented changes in the position of women in western society, also brought what appeared to be scientific confirmation that a woman is something akin to a "failed" man.

I can never quite overcome the suspicion that male teachers of biology often experience a slight *frisson* of self-justification when they recount this story to their students. Male superiority is an ancient fortress, but it has somehow become situated on the new scientific frontier. In a world where men have lost their old birthright, where their confidence has been eroded by women's success in so many aspects of life they once considered their own, is the way in which sex is allocated the last and only true scrap of male supremacy? Female-as-default is a challenging idea because it is politically incorrect but scientifically incontrovertible, at least superficially.

Yet science cannot be used to apply value judgments to its results. If women really are made by default, then does this say anything in particular about women? Also, just because the X chromosome may not carry within it the spark of femininity, this does not belittle its importance. Instead, de Beauvoir recast the whole issue. She pointed out that we are all simply temporary repositories for our genes. Modern biology has taught us that each and every one of us is a vessel containing genes—we get these genes from our parents before they die, and the same genes will be dispersed through future generations after we are gone. The transience of a human life is rather pathetic when compared with the permanence of the genes that life perpetuates. When viewed from this perspective, the actual sex of any one of these evanescent beings seems of trivial importance in comparison to the flow of genes into the future. We are not created so that one sex can be dominant, nor so that it is important how one sex is differentiated from another. All that matters is that two sexes are created, somehow, anyhow, so that our genes can survive.

Having stepped back from the female-as-default orthodoxy, de Beauvoir was then able to attack it head on. Not unreasonably, she suggested that it was simply an arbitrary fact of nature that was being used to support the idea of male superiority. It was certainly not the first biological tidbit that had been used in this way—the actively swimming sperm and the passively waiting egg and the rutting stag and the compliant hind spring to mind. It soon became clear that the misogynistic implications of the female-as-default idea were illusory. After all, if the Y chromosome has to wrest the embryo away from being a girl, does this not imply that being a girl is a dominant state? If I were to assert that all early embryos form with the intention of becoming a girl, it might reverse your view of what being female by default says about women.

So you can interpret the fact that human babies are initially made female, but can be modified into males, however you like. It can mean that women are better or that men are better—it simply depends how you play with the words. Perhaps what all this is telling us is that the discovery of the mechanisms of sex determination has not altered the values we ascribe to the two sexes.

Our conclusion, when all the meaningless philosophical clutter is removed, is fascinating—men and women are made in fundamentally different ways. They are not two sides of the same coin as we always like to think—they are more dissimilar than that. One is the original and one is derived. Rather than futilely trying to apply subjective opinions to that fact, the first thing most scientists would ask is: Does it have to be that way? Is this the only way to make male and female—does there have to be X and Y, a default sex and a derived sex, and does that default sex have to be the female?

There are no simple answers to these questions, but they are worth pursuing for a very important reason. As we try to answer them, we will gradually realize why our sexes are determined in the way they are. Even though we have found the machinery that makes girls different from boys, we will learn much more by finding out why we have ended up with this machinery. We will discover where sex determination came from, and where these little nuggets, X and Y, came into the story. Most of all, we will find out why the X nugget controls the way we live our lives, much more than the Y ever could.

Strong Woman

There is a world of difference between the hormones made by the Y-induced testicle and the XX-permitted ovary. Testicle hormones do things—in short, they turn a baby into a boy. However, there is no real sign that any ovary hormones do much at all to make baby girls girly.

This is not to say that the unborn child's ovaries are not making hormones. In fact, they pump out extremely large amounts of hormones. In many ways, a female human fetus is as reproductively "active" as an adult woman—the baby's ovaries are as fully functional as they will ever be. Yet scientists still think that these ovarian hormones are not actually involved in making an embryo become a girl, although they will certainly be prime movers later in making that girl into a woman at puberty.

To many biologists, this stark difference speaks of something much deeper. The contrasting roles of ovary and testicle, the differing contributions of X and Y, might not be simply arbitrary—they might instead be telling us something. As a result, many have claimed that this apparent asymmetry between male and female is directly related to another great inequality in human life: pregnancy. It may seem a rather obvious thing to say, but it is always women who get pregnant. Of course, this is true of all mammals and you never hear of pregnant male lions, mice, or kangaroos. Although this neat rule rather falls apart outside the mammals—pregnant male seahorses are one example—in mammals the rule always works. Ever since we became mammals, we have all been nurtured inside our mothers.

This is the inequality between the sexes that many biologists claim has led to the way our sex is determined. Each one of us is created, molded, and nurtured inside a female parent. We all develop in a female environment, and as a result, every developing human embryo is awash with potent hormones made by its mother's ovaries. Most important of all, it is awash with these hormones at the very time it is trying to decide its sex. Of course, this is not a problem for embryos that nature intends should be girls—a little more feminine input from their mothers is not going to change matters much. However, for boys, this continual barrage of female hormones presents something of a problem. Despite their feminizing environment, their conversion to maleness must be decisive and complete.

This may be why we maternally nurtured mammals have such an active, strident process that turns us into males. Our ancestors developed in eggs, but as we switched from this method to developing inside our mothers, the pressure was on to find a fail-safe way to make half of us boys. Possibly the Y chromosome and its all-important *Sry* gene were that fail-safe method—a gene that will drive an embryo to maleness, no matter what hormones its mother throws at it. So mammalian embryos became exquisitely sensitive to *Sry* and the hormones made by the testicle, and they became indifferent to female hormones washing in from their mother.

Insensitivity to female hormones is not a problem for mammalian embryos that are destined to be female. The system appears to have evolved so that these embryos will become female by default, simply because they do not have *Sry.* There was simply no pressure to design a fail-safe mechanism to make females, as the maternal environment was likely, if anything, to push them toward femaleness anyway. So while maleness must be forcibly imposed on a baby, femaleness just happens.

I must admit that I do find this a neat idea. It is appealing because it makes some sense, and also it does not seem to have been concocted to make a particular point about the status of men and women. After all, it is beyond argument that women have babies, not men. Although it is a good idea, however, it is far from proven. The problem with theories like this is that they are trying to reconstruct how something evolved, and no matter how good they are, rarely can they be proved right or wrong.

One good way to test such ideas is to try a counterexample, and the obvious counterexample would be to look at what happens when male mammals become pregnant. If we are claiming that our sex determination system evolved to cope with the fact that we are borne by women, what happens when we are borne by men? One obvious problem with this approach is that male mammals cannot (at present) be made pregnant, and so this is an impossible experiment to do. It would be interesting—although ethically dubious—to see what would happen if a man could be made pregnant with an XX embryo. Would the child be masculinized by the male hormones leaching in from its parent? It is certainly not inconceivable—if you will excuse the pun—that such an experiment will be tried within this century. Yet even if male pregnancy does masculinize XX babies, it will still not mean that early mammalian females were at serious risk of feminizing their hoped-to-be-male babies, nor that this was what drove the evolution of the male-directed system of choosing our sex. So this theory remains speculative—like all the most interesting theories.

One thing has become clear over the last decade. The idea that making boys is an active process and making girls is a passive one is not be quite as clear as it had seemed. Indeed, as the chain connecting the Y chromosome to maleness has been slowly revealed, certain findings have cast doubt on the idea that the simple lack of that chain is what makes a girl.

The first sign that things might not be as simple as they seemed came as researchers looked more closely at the *Dax1* gene. *Dax1* is on that list of genes that interact with *Sry* in its attempts to make a testicle. Some of those genes are later links in the chain, and others just seem to help the process along. *Dax1* is rather different. Instead of encouraging *Sry* to make a testicle, it seems to try to hinder it.

At first sight, this did not seem too strange to most biologists. After all, almost every biological system has to be controlled in some way, and most have mechanisms to stop them running amok. It was entirely possible that *Dax1* is just such a control mechanism—a bit like a thermostat to make sure the *Sry* system does not run too "hot." However, evidence is slowly emerging that *Dax1* is rather more strange than that. Most intriguing is what happens when extra copies of *Dax1* are inserted into developing mouse embryos. XX embryos with extra *Dax1* look pretty normal, but XY embryos with more *Dax1* look female.

This is a weirder result than it might appear. We had become convinced that having a Y chromosome with *Sry* on it was all that was required to make us male—think of the smokers in that unusual European restaurant. It did not seem to matter how many X's or Y's an embryo had—any Y's and it would be male. But now we have a gene that spoils this neat story. And worse than that, *Dax1* is not some strange alien gene that messes up the whole of the embryo's development—it is a perfectly normal gene that takes an active part in deciding our sex. All we have done is insert some extra copies of it into mouse embryos.

Dax1 is more than just a way of modulating the male-making activity of *Sry;* it is actually anti-male. It is a factor that actively promotes the development of ovaries and female babies over the development of testicles and boys. Yet it is not the switch that makes babies into girls, because I have already explained that there is no such switch. What this probably means is that there are forces at work waiting to actively, aggressively drive the development of ovaries, should *Sry* fail in its mission. Perhaps de Beauvoir's suggestion that the female is the dominant design of the human body which must be overcome to make boys is turning out to be correct.

A compelling addition to the *Dax1* story comes from humans, and it has two interwoven strands. Apart from *Sry,* all the other genes involved in sex determination are scattered over many different chromosomes—only the master switch has to be on the Y. Rather wonderfully, *Dax1* has been found to be located on the X chromosome in humans. Later on in this book, I will look at more genes on the X chromosome, but it is somehow appropriate that the first gene I mention on our beloved X chromosome should be *Dax1,* and here is the reason why. The second strand of this story is that there are signs that humans have a slightly more *laissez faire* approach to deciding their sex than we once thought. There is now evidence of rare XXXY people who look, for all the world, like women. All our rules tell us that these people should be men, because they have a Y chromosome (and *Sry*)—their party of diners includes a smoker. Yet their abundance of X chromosomes seems to be annulling the effect of the Y chromosome. The nonsmokers have formed a ring around their smoking friend so that no one can see his nicotine-stained fingers, and have marched confidently into the nonsmoking area of the restaurant.

It may be that the normally reticent pro-female influences on the X chromosome can challenge the pro-male factors on the Y. And perhaps *Dax1* is one of these pro-female influences. Even though having a Y chromosome almost always switches on testicles and maleness, lurking under the surface is a more equitable system, in which female and male influences vie for supremacy. Although this struggle is rarely evident, it is still there, and I think Democritus would have felt vindicated.

Dax1 may not even be the last pro-female factor that we will discover. Making something as complex as an ovary is a formidable task, and it is not unreasonable to assume that many dedicated genes will be required to pull off this feat of construction. What if we find that a large number of these genes are carried on the X chromosome too? Would this mean that the X chromosome should no longer be considered the sexless foil for the sexy Y? Later in the book I will return to this question (see the second Interlude, "How Sexy Is X?") in an attempt to discover whether the X carries genes particularly involved in reproduction.

Yet although male-female conflict lies hidden within our genes, the fact remains that Y chromosomes and the *Sry* gene they bear almost always decide what sex we are going to be. Evolution has presented us with a *fait accompli*—this system is how we are allocated our sex, and there is nothing we can do about it. But the nagging doubt remains that things may not have to be that way. And if we want to know if this really is the only way of deciding our sex, we can learn a lot from how other animals do the same thing.

The Birds and the Bees

Scientists no longer believe that the human race is the pinnacle of creation. Two centuries of zoological onslaught have hacked our self-importance almost into nothingness. Instead of being the Almighty's favorite piece of handiwork, humans are just another evolving animal trying to make its way in the world. Although some may hanker for the old discredited human supremacy, our new scientific humility has told us far more about ourselves than thousands of years of ecclesiastical speculation.

We are indeed animals, made of exactly the same stuff as other animals, and this is a very useful thing to know. If ever we wonder why we function in a certain fashion, the best way to find out is to look at our furry, scaly, and slimy relatives. To address the specific question of how babies are allocated their sex, we can ask how other animals decide which sex to be, and compare their decision making to our own. For example, if we were to find that using X and Y chromosomes to make two sexes is a thread that runs throughout the entire animal kingdom, then those chromosomes would suddenly take on a very profound biological importance. We would certainly have to find some pretty good reasons to explain why sex can only be determined

in that way. Alternatively, if different animals pick and choose how
they control the sex of their offspring, then X and Y would be rele-
gated to just one option of many—an option that a few groups of animal might have acquired as necessary during their evolution, and can
presumably discard if they no longer need it. So which is it to be—X
and Y as central biological truth, or transient arbitrary option?

On thing does seem very clear: humans use the same methods of
sex allocation as almost all mammals. Whether they nourish their offspring in a pouch or with a placenta, mammals (furry animals with a
backbone who suckle their young) almost all use the X/Y system.
Female mammals get two X's, but embryos with an X and a Y become
males. These males then randomly distribute their X and Y into their
sperm, and these in turn decide the sex of the next generation. So the
human system is very much the "standard mammal system"—two
sexes, determined by the presence or absence of a single gene, *Sry*,
found on one of the two specialized dissimilar sex chromosomes.

However, as soon as zoologists began to look elsewhere in the animal kingdom, they found that few things that swim, crawl, or slither
have ended up with a system of sexuality exactly like ours. The very
first tenet of mammalian reproduction which turns out to be surprisingly optional is sex itself. In short, many different animals simply do
not have discrete male and female sexes at all. This is especially
strange as within the last century scientists thought they had worked
out why sex is such a good thing for us—every time two animals
come together to make a baby, their genes and chromosomes take
part in an invigorating mixing process that helps discard damaged
genes and disseminate helpful ones. We still think this is why we have
sex, and yet there seem to be many animals that cope perfectly well
without it. Indeed, there is a plethora of ways to reproduce.

The most obstinate celibates of the animal kingdom are animals
that either just split into two by binary fission, like the microscopic

Amoeba, or neatly sprout new individuals out of their sides, like *Hydra,* another favorite of school biology teachers. The other asexual creatures are those that procreate by virgin birth, so-called "parthenogenotes." Parthenogenetic species are entirely female, and each female can produce identical daughters without any need for any input from a male animal—they simply conceive spontaneously. The best known of these ultra-matriarchal species is probably the aphid, but examples closer to home include species of lizard and snake that breed by virgin birth.

So sex, in the human sense, is not obligatory. There are certainly many animals that survive without it, even though it is claimed to confer considerable genetic advantages. Perhaps for some, sex is simply not worth all the effort spent finding, befriending, and copulating with a mate. In fact, many animals get the best of both worlds—they reproduce asexually when they need to throw out a few quick clones of themselves, but they revert to sex when they want to spruce up their tattered genes. When you consider animals as a whole, sex is very clearly optional, and if even sex is not necessary, what chance for X and Y?

Even when we discount nature's celibates and focus on the creatures that reproduce by something we would recognize as sex, we soon discover that our ideas of two fixed sexes—male and female—do not really mean much to many animals. There are many sexual species that cannot be subdivided into neat male and female populations, but perhaps the best examples of the plasticity of sexuality are the hermaphrodites. Named after a mythical son of Aphrodite and Hermes who became blended with a nymph, hermaphrodites are animals that possess both ovarian and testicular material at some time in their life. They reproduce sexually, but they do not have clear-cut, mutually exclusive sexes.

Despite the enormous variety of hermaphroditic animals in existence, there are two basic types of hermaphrodite, and these arrange their mixed sexuality in fundamentally different ways. Some of these species can look just like nonhermaphrodites at first glance—a population made up of a mixture of apparently female and apparently male individuals. Yet if they are observed over time, all the individuals change sex at some point in their lives. Usually they begin as one sex and then, at a pre-determined age, or in response to something changing in their environment, they all switch to the other sex. Their gonads have an ovarian part that they use for half their life, and a testicular part that they use for the other half. These "sequential" hermaphrodites are never both male and female at the same time—instead they start as one and change to the other.

The other hermaphrodites are the "synchronous" ones who actually do operate ovarian and testicular activities at the same time. This potentially gives synchronous hermaphrodites the dubious opportunity to fertilize their eggs with their own sperm—something that some roundworms do, for example—but many others go to great lengths to make sure that this does not happen. Some hermaphrodite fish, for example, employ a mixture of almost tantric self-control and admirable relationship equality to ensure that both partners are able to father and mother offspring, without having to self-fertilize. Serranus fish form pairs in which one partner first ejaculates sperm onto the eggs of the other, and then the two switch roles and the previous egg-bearer ejaculates sperm onto its mate's eggs. One cannot help but feel that this offers a whole new perspective on ensuring that one's partner is satisfied.

Whatever the lurid details, one basic fact is clear: even among sexual species, the distinction between female and male is often not as sharp as it is in mammals. So let us now limit our search to animals that usually do have two discrete sexes, like we do—these animals are called dioecious, meaning "in two houses." Do they all decide the sex of their offspring in the way we do—with X and Y and Sry?

From my previous failure to show that the human system is the one true way to reproduce, you might expect that the answer to that question will be "no." And indeed, there are very many animals that do not use genes to decide their sex—they simply use something else instead. At first, it might seem difficult to think of an alternative to genes, but most species whose sex is not driven by genes use something far less arbitrary—they use their surroundings. Instead of having their sex determined at their conception, they wait and see what their environment is like and choose their sex on the basis of that. This so-called "environmental sex determination" has the inherent advantage that animals can select whichever sex they think will give them the best chance of producing offspring in the future.

It may seem rather risky to base your sexuality on some prediction of what the future holds, but for many species this method obviously makes more sense than slavishly following the orders of randomly inherited genes. One of the clearest and most startling examples of this is the strange lifestyle of *Bonellia,* a worm-like sea creature. Although adult *Bonellia* sit rooted to the sea floor, they produce larvae that float freely in the plankton and eventually settle down at some distant site. Although adult *Bonellia* are either male or female, their larvae are neither one nor the other—they are indifferent. More surprising is the way these larvae decide their sex. If a larva comes to rest on the seabed far from others of its kind, then it becomes a female—a full-blown autonomous mud-living wormy thing. However, if a larva lands next to a sedentary female something completely bizarre happens—it becomes a male, but that is not the end of the story. This new male proceeds to burrow into the adjacent female and develops into what is effectively a large parasitic testicle that survives throughout the life of its bride. One female *Bonellia* can play hostess to several of these rather alarming suitors, and their job is to produce the sperm to fertilize her eggs, so she can make more larvae.

Although the private life of *Bonellia* may seem rather unorthodox, its breeding strategy is not really that different from our own. It has two distinct sexes, and both are needed to make offspring. The only important difference is that these creatures choose their sex according to what other *Bonellia* happen to be around—the parasitic testicle part of the story is really only important because it gets them publicity in books like this. The advantages to the worms are obvious: if they land in uninhabited territory, then they become female, knowing that, should a fellow *Bonellia* come along, it can turn into a male to fertilize them. This form of environmental sex determination makes sense, because it optimizes every individual's chances of having babies.

Even animals as closely related to us as fish—which we know are our cousins because they have backbones too—use a method of sex allocation rather like this, although in fish it is dignified with the term "social" sex determination. It is especially popular in polygamous (or more correctly polygynous) fish in which single male fish keep large harems of females. In the harems of the Ottoman sultans, the sultan's death often led to an uncontrollable bloodbath of recrimination and infanticide, but many polygynous fish have a far more civilized system—in response to the death of the patriarch, one of the females simply turns into a male and becomes the harem-keeper. In this way the harem can remain intact and still produce offspring. I accept that these fish are actually sequential hermaphrodites, rather than dioecious, but the principle remains the same—they can decide their sex to enhance their chances of breeding.

Becoming one sex because all the animals around you are the other sex seems eminently sensible, but many of our closer relatives, crocodiles and turtles, use a rather more unexpected form of environmental sex determination. The cue that they use to choose their sex does not seem to make much sense—they use the temperature of the sand in which their eggs are incubated. For each species that uses this "temperature-dependent sex determination," or TSD, there is a single specific egg temperature that will yield equal numbers of baby boy and baby girl crocodiles or turtles. If the eggs develop above this temperature, then one sex will predominate. This high-temperature sex can be either male or female, depending on the species, but if the eggs incubate in cooler sand, the other sex predominates instead. To take this to extremes, if the sand is very cool or very warm then all-male or all-female clutches can hatch, and indeed this is exactly what happens much of the time.

It is difficult to see why this rag-bag assortment of reptiles should use such a strange method of deciding their sex. TSD is not a rare freak of nature, as many reptiles use it, but we still do not know what possible advantages it brings. Nor do we know why it is not used by any birds. One suggestion is that TSD is just a technique of sex determination that some reptiles were left with because they did not manage to evolve a better one. This seems rather unlikely because many of these reptile species are closely related to species that do not use TSD. Instead it appears that they have expressly discarded other methods in favor of TSD.

Another possibility is that TSD evolved because it was somehow advantageous for these species to produce single-sex clutches. Because of the way their sex is determined, it could be argued that there is not much scope for incest in baby crocodiles, as all their siblings are often the same sex as them. Avoiding incest is not just a social nicety—it actually makes genetic sense. Siblings often share damaged genes, so incestuous offspring are extremely prone to genetic diseases because they get a double dose of these genes. Crucially, baby crocodiles and turtles are hardly in a rush to breed: except for humans, parrots, and whales, they are probably the longest-lived vertebrates in existence. They have 50 to 150 years to make babies, so hatching in single-sex clutches is hardly a problem—in fact, it is a positive advantage if it prevents incest. This seems a clever theory, but on further reflection, TSD is a rather drastic way to avoid incest. After all, many animals are born in mixed-sex clutches or litters, but do not immediately begin copulating with their siblings. Baby crocodiles are, like most babies, infertile, and it is hard to see why this alone should not be sufficient to prevent just-out-of-the-egg incestuous indiscretions.

There is one more theory that attempts to explain why these reptiles use incubation temperature to control their sex, and it is probably the most convincing. Perhaps there is a good reason for males to be born at one temperature, and females at another. Temperature dependence may produce stronger infants, be they cool males or warm females. Intriguingly, there is good evidence that incubation temperature has quite dramatic effects on growth, aggressiveness, and even brain structure in some lizards. Perhaps males are simply produced when incubation temperature has already conferred those characteristics on the clutch that would be most beneficial to males, and of course the same could be true for females. By this logic, TSD could allow infants to be matched to their environment, and maybe even to their own bodies, in some way. Such a scheme even allows mother reptiles to have a hand in that matching decision. At its most extreme, this idea has even been interpreted as a way for reptilian mothers to control the sex of their offspring. Although we do not know why they might want to do this, it is obvious that the only individual who has any control over incubation temperature is the mother who chooses where to lay the eggs.

So TSD is a potentially powerful system for attuning offspring to the environment, although admittedly we do not know why some animals might want to use TSD and not others. Long before anyone knew about TSD, Empedocles claimed that a child's sex is not set at conception, but develops later on, depending on the temperature of the womb, or even the weather. We may have scoffed at Empedocles' idea, but now we can see that it was entirely correct—except that he did not realize that he was writing about turtles and crocodiles.

So, some animals have no sex at all, some animals can be either sex, and some animals decide what sex to be on the basis of where they are, how hot they feel, or which of their friends happen to be nearby. Variety may be the spice of life, but the animal kingdom is being almost ostentatious in its ability to think of new ways of making boys and girls. As a method of reproduction, the X and Y chromosome system certainly does not seem too apparent so far. In fact, as yet none of the animals I have mentioned have any sort of gene-based or chromosome-based method of deciding their sexuality. So I will now whittle down our search to animals that do use such a method.

The creatures that use genes or chromosomes for sex determination are a mixed bag, scattered fairly randomly around the animal kingdom. They make up a good fraction of all the animals in existence, and so this method is certainly a mainstream technique for choosing sex. However, the variety of animals with genetic sex determination is also reflected in the diverse ways that they actually use those genes—once again, I will show that a macho Y chromosome imposing its will on an initially female body plan is far from the norm. Active Y and passive X are not consistent themes throughout evolution, and when I describe some gene-sexed animals, you will see how the apparent submissiveness of X to Y may not be real at all.

Some gene-sexed animals simply do not have any sex chromosomes. Instead they use a rather more dramatic trigger to make male and female offspring. Honey bees are a good example of this sort of sex determination. You may remember that Nettie Stevens noticed that chromosomes often come in pairs. Indeed, humans have twenty-two matching pairs of non-sex chromosomes, and of course women

have a pair of similar sex chromosomes too (X and X). This duplication of genetic material is a common feature of many animals—
inheriting a copy of every gene from both of your parents
considerably improves your chances of getting at least one undamaged copy of most genes—but bees also use this duplication to control their sex. Queen bees have two sets of genes arranged in
chromosome pairs, just like you or me. When queens are mated by a
drone bee, each parent then contributes one of each pair of its chromosomes to each baby female worker bee—so worker bees have
duplicated genes and chromosomes like queen bees and human
babies. However, queen bees can make drones without using any
sperm and because of this, drones only get one set of genes. Drones'
chromosomes are not paired, so they have half as many of them as
workers and queens. In fact it is this reduced number of chromosomes that actually makes them male. Hence, in bees it is the number
of sets of chromosomes that decides the sex of offspring, and not the
presence of any particular gene.

So although bees certainly do use their chromosomes to decide
their sex, they do not have recourse to any specialized sex chromosomes. Their sex determination is very different to ours, yet using
chromosome number to control sex is not just some freakish idiosyncrasy of bees—it is widespread throughout the animal kingdom, and
is even used by vertebrates—some fish, for example. Using chromosomal sex determination does not mean that you must have sex
chromosomes.

Whittling down our list further, we come at last to those animals
that have true, dedicated sex chromosomes like us. In all these animals, at least one pair of chromosomes is dissimilar in at least one
sex—just as X and Y are dissimilar in men. Yet even here there are several variations on the sex chromosome theme, and the most important way in which these animals can differ is in the way that their sex

chromosomes actually control their sex. We already know that the human Y chromosome makes embryos into boys because of a single dominant gene called *Sry*. However, there are plenty of animals in which neither sex chromosome carries an all-controlling dominant gene, and instead it is the number of sex chromosomes present that controls sex.

A good example of this is that biological workhorse, the fruit fly *Drosophila*. Fruit flies have big X and little Y sex chromosomes like we do, but they work in a very different way from ours. It is the number of X chromosomes that decides the sex of a baby fly. If a maggot has two X's then it will be female, and if it only has one X, then it will be male. In the normal course of events, this means that fly sexiness can look very like ours—XX females and XY males—but flies with abnormal numbers of sex chromosomes often become a different sex to humans with the equivalent sex chromosome tally.

Fly: XX, female; XY, male; XXY, female; X–, male

Human: XX, female; XY, male; XXY, male; X–, female

This dabbling with intersexual flies may seem a bit esoteric, but it demonstrates a crucial difference between the control of sexuality of flies and people. In people, it is the Y chromosome that imposes maleness on an embryo, whereas in flies the role of the Y is passive. Fly Y is just a hanger-on, a makeweight to fill the missing space in single-X male flies. The human Y is the arbiter of sexuality, but in flies it is the number of X chromosomes that a maggot receives which decides its sex.

If the Y is not really doing anything to determine sex, then you might wonder why is it there at all. And indeed, many insects have dispensed with it altogether—a Y chromosome is just an optional extra when you use X chromosome dosage to determine your sex. In other words, many insects species are made up of XX females and X– males, and this is why many of the early studies of sex chromosomes produced apparently confusing results: it depends on which insect is studied. The flies' way of determining sex is actually quite widespread throughout the animal kingdom. For example, another biological research favorite, the roundworm *Caenorhabditis*, has two sexes—hermaphrodite and male. The former are XX and the latter are X–.

Now, another question raises its ugly head. Are females always the XX sex and males always the X– or XY sex? Perhaps this could be better phrased: Do females always have sex chromosomes that are alike and males always have sex chromosomes that are different (the technical terms are "homogametic" and "heterogametic"). In fact, many sex-chromosome animals do indeed have this arrangement, but we do not have to look far before we find a large group of animals that have the opposite arrangement—we have already considered the bees, and now it is time to look at the birds.

Birds are, intriguingly enough, the other way round from mammals. Cocks have a ZZ chromosome pair, and hens have a ZW chromosome pair. It is almost as if birds are deliberately trying to challenge our views of sex determination, and make a place for themselves on the pages of biology textbooks. After all, you may argue that the sexual proclivities of roundworms are of limited interest to the general public, but everyone knows what a bird is. Birds have turned out to be resolutely different from us, and yet unfortunately they do not seem keen to divulge their sex-determination secrets.

The first thing we would like to know about bird sex chromosomes is whether they act by a single dominant gene, like ours, or if they work by a dosage system, like flies. In other words, does the lone W make a chick into a female, or do two Z's make a male? I have already mentioned how individuals with abnormal numbers of sex chromosomes have helped us answer exactly this question in humans and flies. Now geneticists need to find out what birds with unusual sex chromosome complements look like. They have spent a rather endearing amount of time trying to find ZZW and Z– birds. Yet their search has been frustrated by the extreme rarity of abnormal sex chromosome birds—some sex chromosome combinations have never been detected in birds. No one knows why this is—perhaps they all die before they hatch—but it has demonstrated just how important sex chromosome abnormalities have been in divulging how human sex determination works. So the jury is still out on how birds decide their sex. Next time a bird perches outside your window, just remember what an enigmatic fellow it is—is it the opposite of a fly, or a weird sort of reverse-person?

By the wonders of modern genetics, we can now compare mammalian X and Y with avian Z and W sex chromosomes, and they have turned out to be strikingly different. The sex chromosomes of furred and feathered vertebrates do not appear to have very much in common—no large tracts of shared sexy genes—nothing. In fact, large chunks of mammalian X and Y are actually very similar to regions of the non-sex chromosomes of chickens. Because of this, scientists now think that mammals and birds have not taken the same sex chromosomes and simply used them in slightly different ways—instead they have independently constructed their own distinctive and completely unrelated sex chromosomes.

It is really rather remarkable that the two most successful groups of contemporary land vertebrates use methods of sex determination that are so very clearly unrelated. This strange difference is thrown into sharp relief when you consider that a few hundred million years ago, a little reptilish thing was scuttling around which was the ancestor of both mammals and birds. The two groups of animals may seem very different, but they must have evolved from some common ancestor in the dim and distant past. And if these descendants use methods of sex determination that are almost diametrically opposite, then how did that scuttling ancestor determine its own sex? This question becomes even more challenging when you consider that that creature was also the direct ancestor of snakes, lizards, and crocodiles—animals that use a variety of chromosomal and environmental sex determination methods, and that occasionally even reproduce by virgin birth. It is easy to be fooled into considering only the animals alive today and how they decide their sex, but they all evolved from extinct animals that must also have chosen to be male or female in one way or another. We would dearly love to know how animals can evolve to use new sex determination systems and discard old ones, all the while remaining fertile, breeding, two-sexed populations. We would love to know, but we do not, and the bird/mammal mystery remains unsolved.

And now at last I come to the mammals. In the morass of sex determination that is the animal kingdom, other mammals seem reassuringly like us—XX females and XY males, and that is all there is to it. Though other creatures may have spurned the X and Y that are the central tenets of human sexual ideology, our fellow mammals seem happy to conform to our expectations at last.

Except that they do not. Even in mammalian sexuality, all is not as simple as it seems. As if all the other strange beasts were not enough to demonstrate that the X and Y chromosomes are mammalian idiosyncrasies, ephemeral minutiae in the grand flow of evolution, evidence is coming to light that even mammals may not be as uniformly committed to the X/Y/*Sry* system as scientists had thought.

One problem is the egg-laying mammals or "monotremes." There are very few of these, but they include those exotic but extremely cute Australians, the duck-billed platypus and spiny anteater, or echidna. Whenever a biologist tries to make a general statement about mammals, it is always the monotremes that will prove them wrong. Few they may be, but this distantly related group are very much mammals—they suckle their young just like us. And while platypus and echidna have X and Y chromosomes, they do not seem to have *Sry*. This may be because their version of *Sry* is so bizarre and archaic that geneticists have simply failed to identify it, or more likely it really is not there at all. Their likely lack of *Sry* has probably put monotremes outside the standard mammalian system of sex determination. Maybe they control their sex by their number of X chromosomes instead—a dosage system like the fly. We simply do not know.

Whereas monotremes can always be argued to be a minority interest, marsupials in contrast are very much mainstream mammals. Instead of rearing their young to relative maturity in the womb, they give birth to nude, immature offspring, but then feed them up on milk in the protection of a pouch. This pouch is what defines this large group of Australian and American creatures, but it is also the pouch that betrays a subtle difference between them and us. Marsupials, as far as we know, use the X/Y/*Sry* system to decide their sex—the presence of a *Sry*-bearing Y chromosome initiates the construction of testicles, sperm tubes, and penis, just as it does in humans. Yet there is one exception to this rule, and that is the pouch. Marsupial embryos have a bulge on their belly, and in females this becomes the pouch, whereas in males it becomes the scrotum—the sac in which the testicles will dangle. However, this is what happens in marsupials with normal sets of sex chromosomes. In XXY marsupials, the Y chromosome drives the formation of most of the male anatomy, but instead of a scrotum, these confused individuals grow a pouch. Scientists now think that this pouch forms because they have two X chromosomes, like a female—even though no other part of their body ever becomes feminized. So in a quaintly pouchy way, marsupials have two distinct methods of sex determination: human-like *Sry*-controlled testicles, tubes, and penis, but a flylike X chromosome dosage-dependent pouch.

The other main thread of mammalian evolution is us "placental" (or more correctly, "eutherian") mammals—pouchless furry animals who nourish our offspring in the womb with a well-developed placenta. With few exceptions, all mammals outside Australia are placental mammals, including humans. Here at last is a homogeneous group of animals that all use the standard mammalian system of sex determination—people, pets, farm animals, mammals that swim, mammals that fly, and mammals that drift majestically across the plains—wherever one looks, *Sry* retains its vice-like hold over our sexual destiny.

Or so we thought. In biology there is always an exception that disproves the rule, and the exception that disproves this one is the humble mole vole. You have probably never heard of mole voles, and I must admit that I had not heard of them either before I started researching this book. They are, by all accounts, voles that look slightly like moles, yet just like *Bonellia* and its parasitic testicular husbands, mole voles will forever be writ large in the annals of reproductive biology. Zoologists could not have dreamt up a better sexual oddity. Some species of mole voles are just like "normal" mammals— they have XX females and XY males. But other mole vole varieties have XX females and X– males—they have lost the Y chromosome and they have lost *Sry* along with it.

The mole vole is a far more profound challenge to the supremacy of the standard mammalian system of sex determination than the undecided monotremes or marsupial pouches. Alone among the myriad ranks of placental mammals, mole voles have broken ranks and broken free from the tyranny of the Y chromosome. It is almost as if they are the only placental mammals to have realized that they do not need to have an *Sry* gene at all, and maybe they now decide their sex by X chromosome dosage instead.

Mole voles' bold schism with the rest of the mammals leaves our X/Y/*Sry* system of sex determination in a rather strange philosophical position. It would be easy to accept that all placental mammals use the same system—presumably the ancestral mammal had this system, and all its descendants simply inherited it. Equally, we could understand if mammals fell into two large groups, with one group using one method, and the other using another—this would simply mean that mammals could select one of two systems according to their needs. Yet the situation in the real world is really rather inexplicable—the vast majority of mammals use the X/Y/*Sry* system, as if this is the only possible method, and yet a tiny minority of an obscure subgroup of voles shows that this is very clearly not the case. Why should mole voles be so different? It would not be surprising if bats and whales were different in some way, but why different types of vole? I wish I could tell you why these small furry creatures have taken it upon themselves to break with mammalian sexual orthodoxy, but I cannot.

If there is one thing that our tour of the sex life of the animal kingdom has told us, it is that there is nothing special about the way that the sex of human babies is decided. It is certainly not the only sex determination method in town, and indeed it sometimes seems as if nature has gone out of her way to invent new ways of making girls different from boys. Neither the X nor the Y chromosome is cast in stone—both are optional entities that animals can acquire and discard as they evolve. The mole voles have shown us that even among the mammals there is evidence that the X/Y/*Sry* system can be cast aside in favor of a new system.

Yet although they are transient and disposable in the evolutionary scheme of things, the X and Y chromosomes are what we use here and now. Whether or not flies or worms have equally clever systems, we humans have to live with dear old X and Y. Thrust upon us by the vagaries of evolution, they hold great sway over our lives, and yet only very recently have we found out where we got them. All our other chromosomes come in neat pairs, but men inherit a messy pair of unequal sex chromosomes—gangly big X and stunted little Y. Scientists now think they know whence this odd couple came, and what they have learned has shown us that the two are not perhaps as different as they may seem.

Drifting Apart—the Sad Divorce of X and Y

One thing about the X and Y chromosomes seems very clear. They are relatively recent inventions. They may be pivotal in controlling that all-important characteristic—our sex—but that importance is not reflected in their evolutionary antiquity.

All animals arrange their genes onto chromosomes for easy storage, and as we have seen, each individual often inherits two copies of each chromosome—one from each parent. Because chromosomes are so ubiquitous, it is likely that we evolved them a very long time ago. In fact, plants have chromosomes too, so they probably date from the almost unimaginably distant epoch when the common ancestor of plants and animals swam the primordial seas (I will leave it to you to speculate about what this strange individual actually looked like).

However, there is a great deal of evidence to suggest that our sex chromosomes are not nearly as old. That scuttling common ancestor of mammals, birds, and modern reptiles that we met earlier probably did not have X and Y, and only lived about 300 million years ago, a mere moment in evolutionary history. So mammals probably evolved their X and Y in the last 300 million years, and of course birds also probably evolved their W and Z within that same period.

Apart from their relative novelty, the other thing that scientists have learned about X and Y from other animals is that mammals probably cobbled them together from non-sex chromosomes. Not only do they look suspiciously like some of the non-sex chromosomes of chickens, but also, if we did not make them from non-sex chromosomes, then it is difficult to see what else we could have made them from. When all is said and done, X and Y are just chromosomes, albeit strange ones, and they must have been fashioned from other chromosomes in the distant past.

Within the last thirty years, genetic science has provided compelling evidence that our X and Y chromosomes originally evolved from a matching pair of non-sex chromosomes. They may now seem as different as chromosomes can be, but their differences belie their common origin. Although the sex chromosomes were originally discovered because they do not take part in the chromosomal dance that takes place when sperm are made, they are not entirely averse to meeting and mixing. There is one small region of the Y chromosome that occasionally swaps with an equivalent region on the X chromosome just before sperm are produced in the testicle. Because this region is often swapped between X and Y, the two chromosomes are extremely similar in this area—far more similar than they are over the rest of their length. In fact, this small region acts very much like a non-sex chromosome, as it can happily swap between chromosomes, and so I will call it the "non-sexlike region" (this was the best name I could think of: geneticists call it the even more impenetrable "pseudo-autosomal region").

The non-sexlike region really is very like a piece of a non-sex chromosome: it carries lots of genes that do useful things, and it is constantly refreshed by bouts of gene shuffling between X and Y chromosomes, and indeed between X chromosomes as well. It cannot really be expected to carry genes that control the sex of babies,

because it is constantly being shunted between X and Y chromosomes, and that would lead to all sorts of problems.

Yet the non-sexlike region helped us to discover the *Sry* gene.
Because *Sry* controls our sex, it is supposed to stay on the Y chromosome, and so it is not in the non-sexlike region. However, for some
reason that no one can explain, *Sry* is located perilously close to that
region, and because of this it can be accidentally swapped onto the X
chromosome when sperm are made in the testicle. This rather sloppy
approach to chromosome splicing is what occasionally leads to the
birth of XX boys. Conversely, when *Sry* is accidentally lopped off the
Y, this can lead to the birth of XY girls. You will remember that it was
by studying XX men and XY women that geneticists actually discovered *Sry* in the first place. In other words, were it not for the apparently daft location of *Sry* on the Y chromosome and sperm cells'
apparent inability to excise the non-sexlike region accurately, we
might not have discovered the very gene that turns human embryos
into boys.

So there is every reason to believe that our X and Y chromosomes
started life as common non-sex chromosomes, and scientists now
believe that they can piece together the story of how this happened.
Probably the first step was that the two chromosomes stopped swapping genes over most of their length. The couple first stopped dancing, and then they almost stopped communicating completely. They
kept in touch by way of the non-sexlike region, but this was not really
enough to stop them drifting apart.

When couples look back over a divorce, it can be hard to remember
exactly when different parts of the relationship started to change, and
the same is true of the chromosomal divorce of X and Y. A crucial
moment in the story of X and Y was when the sex-determining gene
appeared on Y, but scientists still argue about whether this occurred
before or after the two chromosomes stopped taking to each other. For

example, the presence of this gene could have been exactly the trigger that encouraged the chromosomes to stop swapping genes—then one chromosome would carry the gene (Y) and the other would not (X).

Yet others who have studied the evolution of *Sry* claim that the opposite is true and that the break between X and Y was irrevocable before *Sry* appeared on Y. Remarkably, scientists think they may have identified the gene from which *Sry* may have evolved, and it is challengingly situated on the X chromosome. It is called *Sox3* (a completely different gene from *Sox9*) and it has several similarities to *Sry* that suggest that the two have a common origin. At first sight, this might make it seem as if the ancestor of *Sox3* and *Sry* sat on the non-sex chromosomal ancestors of X and Y, and that the lack of communication between the two chromosomes led to one copy of the gene changing into *Sry,* the gene that controls our sexuality.

However, evidence that the lack of communication between X and Y may have predated the arrival of *Sox3* comes from a strange quarter. In egg-laying mammals—echidna and platypus—we have already seen that there does not appear to be a *Sry* gene, but what about *Sox3?* Well, *Sox3* is there, but it appears to be in the "wrong" place—it is on a non-sex chromosome. So here we have mammals with X and Y chromosomes, but with no *Sry* and no evidence that *Sry*'s partner, *Sox3,* has ever been located on a sex chromosome. This seems to be pretty good evidence that, contrary to what scientists thought at first, the X and Y chromosomes were leading separate lives long before one of them acquired the gene that was to become the master controller of sex.

Regardless of the chronology of their divorce, the separation of X and Y was to change them both forever, and to change them in very different ways. When it acquired the sex-determining gene, the Y chromosome consigned itself to a rather grim future. Except for the non-sexlike region, it cannot exchange genes with the X chromosome, and it can never exchange genes with other Y chromosomes because it almost never coexists in the same cell as another Y. So after its separation from X, the Y chromosome became very isolated and started to lose its grip on life. It could not repair its genes by interacting with other chromosomes, and so most of its genes either fled to other chromosomes or simply deteriorated into nothingness.

Because of this, most of the Y chromosome is a wasteland, full of junk fragments of damaged genes interspersed with a few genes that have managed to cling on through the bad times. The only exception to this bleak picture is, of course, *Sry,* the gene that consigned the Y chromosome to this fate. So, apart from the non-sexlike region, the Y chromosome is really just a vehicle to carry the *Sry* gene—a stunted, damaged, introverted shadow of its former self that is so obsessed with controlling sex that it has become almost incapable of doing anything else. Apart from *Sry,* the Y chromosome has become almost useless, and indeed many marsupial mammals show their disdain for it by discarding it from most cells in their body. Likewise, I have already mentioned that mole voles have unceremoniously dumped it, with no apparent ill effects.

There is a good reason why the Y chromosome cannot contain many important genes: females have to spend their whole lives without it. The near-emptiness of the Y probably explains why babies born with extra Y chromosomes do not seem very abnormal. Usually, inheriting an extra chromosome is either lethal, or it affects babies quite dramatically, as in Down syndrome. These effects probably result from the baby being overwhelmed by the excess genes on the extra chromosome. However, there are very few genes on Y that could overwhelm an XYY, XYYY or XYYYY baby, and this probably explains why these boys are only distinguished by being slightly taller, having more learning disabilities, and, controversially, perhaps being more inclined to criminality. Inheriting extra Y chromosomes is hardly a death sentence (except sometimes in Texas).

After the X and Y chromosomes went their separate ways, the X fared much better than its old partner. The main reason for this was that it was much more sociable—every time eggs were being made inside mammalian ovaries, every X chromosome was able to spruce itself up by swapping genes with its fellow X chromosome. Unlike the Y, the X has not committed itself to a hermitic life, and in every female mammal X chromosomes can interact freely with each other, just like non-sex chromosomes. Because of this, the X has not suffered the degeneration that has blighted the Y. I will explain later in this book how the X may have become slightly specialized, but in general it is still a fully functional chromosome, just like any other.

This is why the X is a full-size, apparently normal chromosome, and the Y is a tiny shrunken waif. And of course the X chromosome is still able to carry genes that are essential for life, since both males and females inherit X chromosomes. The X is flourishing, and the reason for this is that its main role is not just to carry a single gene. Rather than being the passive counterpart of Y, as we once thought, X is the partner that emerged unscathed from the sad divorce of X and Y.

The human sex chromosomes are an odd couple. Plucked from among the other chromosomes, they started as a conventional pair, but once Y acquired the ability to control the sex of children, it was destined to fade into a shadow of its partner, the X chromosome. As a result, the Y chromosome is a single-issue chromosome, whereas the X chromosome controls our lives in thousands of different ways. That is why this book is about X chromosomes and not Y chromosomes. If I had decided to write about the Y chromosome, then I would have to stop here, as this is where Y's story ends.

In contrast, the X chromosome has ended up being the most interesting chromosome of all, even though it looks superficially like all the non-sex chromosomes. The reason that the X controls our lives is that it has no real partner, and this causes all sorts of problems in both men and women. It is as if human sexuality used to be a two-party democracy, but one party (Y) became so obsessed with a single issue (sex) that it sank into obscurity. The X chromosome has been left with no effective opposition, and has become our dictator.

The X and Y chromosomes may be just a transient evolutionary whim, but they are central to our lives right now. Especially, we depend on our X chromosome as on no other chromosome. The whole of human biology seems to be designed specifically to tackle the two problems the X chromosome has caused. First, how do men cope with just one X chromosome? Second, how do women cope with two? These are not trivial questions: any animal that fails to answer them is doomed. The other two chapters in this book will investigate how we deal with these problems. The Y chromosome may determine our sex, but the X determines whether we live at all. This is no longer just a matter of sex—it is a matter of survival.

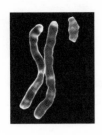

INTERLUDE:
WHAT IS IT, EXACTLY?

So, what is an X chromosome?

It is impossible to avoid genes nowadays. The media are crammed with news about genetic engineering, genetic modification, or gene therapy. Yet very rarely does anyone try to explain what these genes are and what they do, or try to reconcile the concept of the gene with those other two *bêtes noires* of the jargon-averse: DNA and chromosomes. So that is what I will now try to do.

Probably the least impenetrable way to explain genes, chromosomes, and DNA is to approach them in historical order. An understanding of genes came first, in Moravia, with the publication of Gregor Mendel's work in 1866 on inheritance in peas (although the significance of his work was not widely appreciated for many years). His experiments appeared to show that characteristics are inherited by physical means, by the bequeathing of sets of instructions by parents to their offspring. No one at the time knew what these instructions actually were, but they were soon given the name "genes" anyway.

Next came chromosomes, around 1880 in Kiel, Germany. Walther Flemming first described how the dark-stained material in the cell's nucleus condenses into tiny threads each time a cell divides. He counted the same number of threads in every cell in an individual, and in fact any member of the same species, and also gave them the name "colored bodies," or "chromosomes." It had long been claimed that hereditary information is carried in the cell's nucleus, so it was not long before scientists started to wonder whether genes were carried on chromosomes.

The story now pauses for several decades and leaps from Mitteleuropa to Cambridge, England, and DNA, just after the Second World War. For some years before James Watson met Francis Crick, it had been known that the cell nucleus contains surprisingly large amounts of an acid, deoxyribonucleic acid, or DNA. Watson and Crick were both fired by the belief that DNA was the stuff that contained genes, and passed them on to the next generation. However, they had no good evidence for this belief, until they attempted to work out the structure of DNA from extremely pure samples made by Rosalind Franklin and Maurice Wilkins. To cut a very long story short, DNA turned out to have a rather delightful design. It is a very very long molecule made up of two strands connected along their length like a ladder, and as if to make it more attractive, this ladder is wound up into a helix—a sort of ladder-*cum*-spiral staircase—the "double helix" that has made its way into scientific folklore. Watson and Crick were quick to notice that the rungs of the ladder could be varied to create a code which might carry the instructions that make up genes. Also, to use their own charmingly understated phrase from their 1953 paper, "it did not escape their notice" that the two sides of the ladder could be pulled apart, but each would retain the code, so that each could be used to reconstruct a perfect copy of the original ladder—providing a possible means by which genetic material could be replicated.

So there, in three paragraphs, you have it. Our individual characteristics are created largely by the action of genes, which are held on long linear molecules of DNA. These molecules of DNA are, in turn, packaged into chromosomes, visible through a microscope in the nuclei of dividing cells. That DNA is peeled into two halves every time a cell divides, and so every chromosome must be teased into two daughter chromosomes at the same time. This peeling also takes place when eggs and sperm are made, and this is how genes are passed to offspring. And since 1953, biologists have been sorting out the details of how all this works.

Most important, we now know what those genes actually do. Most of them provide codes for making proteins, and proteins do just about everything that a cell does. Some proteins make the structural components that support and move cells, and indeed our bodies. Other proteins are called enzymes, and they spend their time chopping and changing other molecules, including nutrients, waste, other proteins, and DNA itself. Proteins are so important that now that the Human Genome Project has deciphered the entire code of all the DNA in all the human chromosomes, many scientists believe that the next obvious step is to hunt down all the proteins that they make.

So this little history lesson tells us a great deal more about the X chromosome. Rather than being an ill-defined splodge of stuff that somehow came to control our lives, it is now revealed as a very long ladder-shaped molecule, and its rungs carry the code for lots of genes.

A common misconception abut the X and Y chromosomes is that they were named because of their shapes. However, we have already seen that Hermann Henking named the X because it was mysterious, or exceptional in some way, and that Nettie Stevens then simply continued the alphabetical sequence when she named the Y. To be honest, most of the time the X and Y chromosomes are very fuzzy in appearance, and do not look like any letter of the alphabet. But when cells divide, all the chromosomes condense down, and transiently

look like chunky little cruciform things, with a blob in the middle. As
a result, the X and Y look like this:

The X chromosome takes on an X shape, but this is hardly a defining
feature because at this point all the non-sex chromosomes look
X-shaped as well. However, the Y, although it has the same basic
structure, is so small that two of its four arms are often indistinguishable, giving it an apparent Y shape under the microscope. Whatever
their appearance, very soon all these X-shaped chromosomes are
torn apart, and each daughter cell receives only half of each of them:

Then, the chromosomes "decondense" back to their usual indistinct
fuzzy appearance. So the X and Y chromosomes resemble their
namesakes only briefly—one of the most unlikely and confusing
coincidences in all of science.

   Recently, we have found out how many rungs the X chromosome
ladder has—approximately 160 million. Because of this we can set
about working out how long an X chromosome is, as we know that

one rung of DNA is 332 trillionths of a meter long. So, by simple multiplication, we can calculate that the DNA in an X chromosome is 53 millimeters long, or just over two inches.

Two inches may not seem very much for something that carries 160 million pieces of information, but it is very long indeed for something that has to be crammed into every cell in the human body. Fortunately, DNA is not only very long, but it is also very thin, and the reason that the X chromosome is not two inches long is that DNA can be folded. The DNA is first wrapped around special packing proteins, and these are then repeatedly packaged and repackaged until all that long spindly DNA can fit into a single cell. This supreme feat of folding results in a stubby little strand of material called an X chromosome, about eight millionths of a meter long.

But is length really that important? Perhaps instead we should work out how much an X chromosome weighs. Luckily, once again, the world of genetic trivia comes to our rescue. We know that a single rung of DNA weighs 1,054 septillionths of a gram. That means that an X chromosome weighs 169 quadrillionths of a gram, which is considerably more, but still not very heavy. In fact, it gives you an idea of how very thin DNA is when you think that the DNA in an X chromosome is two inches long, but only weighs 169 quadrillionths of a gram.

Anyway, this leads us on to our next feat of biologico-mathematical trivia. If we know how much an X chromosome weighs, can we work out how much of the human body is made of X chromosomes? This, unfortunately, is the first time that we will have to make a guess. No one can agree on how many cells there are in the human body, as no one can be bothered to sit down and count them. I would say that a good estimate is 20 trillion, but please do not ask me why. This would mean that the average male human would contain 3.3 grams of X chromosomes, or just over one tenth of an old-fashioned ounce, and of course an XX woman would have twice as much X chromosome in her. This means that about .005 percent of a man is made up of X chromosomes, and about .01 percent of a woman is X chromosomes.

So, in brief, the X chromosome that controls our lives is a two-inch-long string of code that is wrapped up tight, is not usually X-shaped, and is unlikely to form a significant part of any weight-loss program.

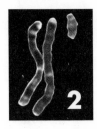

2

THE DUKE OF KENT'S
TESTICLES

Edward paced unhappily around the library in his nightclothes. It was three o'clock in the morning but he simply could not sleep. That July night in 1818 was unusually humid, and he had long since given up trying to rest.

Besides, he rather disliked the city and the oppressive sense of duty that always seemed to weigh on his shoulders when he was there. Edward Augustus Hannover, Duke of Kent, was always far more content away from London, and this enforced stay at Kensington Palace at the height of summer was becoming unbearable. He had never wanted to marry, and had happily reached the age of fifty before giving in. But now the deed was done—a few days ago—and he could look forward to spending the money that had been his inducement to marry.

He did not dislike his new wife, other than that she was a wife and he had never wanted one. Luise Viktoria of Saxe-Coburg was pleasant enough, and despite having been through the marriage mill once before, she seemed fairly keen to carry out her constitutional duty— to produce a child of some description. The plan seemed to be working for now, and indeed for all Edward knew until his dying day, it worked perfectly. Yet the agent of that plan's destruction was at that very minute approaching the top of Edward's head at alarming speed.

It had all been rather unfortunate, but Edward could hardly blame his older brothers. They had simply tried to follow their hearts the same as him, and yet between them they had brought the British monarchy to the point of extinction. Though their father, George III, was still alive, the nightmare of his madness still cast a shadow over the whole family. For some unfathomable reason, his illness seemed to have made him more popular than ever with the people, yet it had made his sons' lives ever more grim. Edward's oldest brother, George, the Prince Regent, was hardly the best rock on which to build a new dynasty. He had not endeared himself to the populace or the establishment when in his early twenties he "married" a twice-widowed Roman Catholic commoner, Maria Fitzherbert. Young George's obsession with her had been a constant thorn in his father's side, since an act of Parliament specifically excluded any royal marriage with the slightest whiff of papism. The marriage of 1785 was thus invalid, and so any children it yielded would be unable to inherit the throne.

Eventually, George's parents forced him to enter a more acceptable union, this time with the rather dull Caroline of Brunswick in 1795. Although dullness was the characteristic least likely to endear a bride to the vivacious, hard-living George, Caroline was exactly what his parents had wanted—an unequivocally genuine Protestant. And German too. George rapidly came to loathe her, later tried to divorce her, and was even soon to attempt to exclude her from his coronation as George IV. Yet before their relationship broke down entirely, in the first year of their marriage George had managed to stir his ever-increasing corpulence to impregnate her with a child that was apparently the solution to everyone's problems. Princess Charlotte was born in 1796 and to George's absolute delight she turned out to be just like him—a hedonist who rather tired of her mother's uninspiring personality. Everything was looking bright when this queen-in-waiting married a German prince, and quickly became pregnant. Yet hopes of a direct royal succession collapsed when she died in child-birth at the age of twenty-one, and her baby was buried with her.

This traumatic episode had played itself out the previous year, and it was remarkable how Edward had now found himself ensnared in this rush for an heir. Both he and William, the second oldest brother, had been pressured into marrying that year. William's personal life had drifted along rather enjoyably until it was thrown into turmoil by the death of Charlotte. For twenty happy years he had lived, un-married, with the actress Dorothy Jordan at Bushey Park in London, and their ten children were testament to their mutual attraction. Their children's illegitimacy was no great problem in itself, until William was required to produce a true heir. William married a German princess the same year as Edward—Adelaide of Saxe-Meiningen—and yet tragedy was soon to strike when their two daughters died in infancy. William was to become William IV for a few years, but he was to leave no heir.

So as Edward paced the room, he justifiably wondered how he, George III's fourth son, had become such a central part of the great drive for the survival of the monarchy. Prince George's beloved daughter was dead and her baby with her, the Regent himself was rap-idly becoming incapable of siring a child, especially with his hated wife, and William's babies were doomed. And now Edward's silent nemesis entered the room.

Imagine that millions of years earlier, a distant star reached the end of its life. Many times larger than the sun, it could not simply fade away into obscurity when its fuel was burnt. Instead, its rapid collapse was followed by an even more dramatic rebound explosion, a super-nova. In a stellar convulsion of almost inconceivable violence, much of the material of the old star was launched into space at incredible speeds—many atomic fragments were accelerated almost to the speed of light.

Of course, Edward had no concept of what a cosmic ray might be, nor that one might blight his descendants for the next four generations and change the course of world history into the bargain. Indeed, the impact was painless. The particle may have traveled from outer space to Edward's testicle in less time than it takes to blink an eye, but he was entirely unaware of absorbing its momentum. History has not been gracious enough to inform us which testicle was the target, but we do know what happened when the ray hit Edward's crown jewels. Like most men, Edward's testicles were a ferment of sperm-producing activity, and they were especially active at the moment, what with his recent wedding. The cosmic ray wrought its greatest destruction on a small cell huddled against the wall of one of the thousands of tiny sperm-producing tubes that filled his testicle.

As it fizzed through the cell, the cosmic ray neatly punched a hole in the cell's X chromosome. The particle was long gone by the time the cell's repair machinery started to tidy up the damage. Hundreds of specialized proteins swarmed around the blighted chromosome and glued it back together, but there was one region that was so disrupted that there was no trace of the genetic code that had previously been there. So the cell had no choice but to patch up this region as best it could. At last Edward retired to bed, but the damaged cell was beginning to divide. Over the next few weeks it would produce four sperm, two of which would carry copies of the damaged X chromosome, and one of those would eventually find itself halfway up Luise Viktoria's Fallopian tube, fighting its way into an egg. This unusual convergence of monarchy, supernova, and testicle set the scene for royal genetics for the next century.

Born the next year on the twenty-fourth of May in the same palace, the baby's parents had wanted to call her Georgiana, in an echo of her uncle, grandfather, great-grandfather, and great-great-grandfather. But her uncle George did not like this idea at all, and his word was law, so instead the child was called Alexandrina Victoria. The birth of this healthy little girl was a spark of hope that punctuated a long dark decline for the British monarchy. Edward Hannover died early the next year, 1820, but not because of any testicular misfortune. He died of pneumonia contracted after a winter cold. Within a week, Victoria's grandfather George III also died—mad, sad, but popular.

George IV lurched onto the throne, but the people's enthusiasm for the monarchy deteriorated along with George's own physique and moral fiber. By 1830 he too was dead, leaving the throne to his child-less sixty-five-year-old brother William IV. Whereas George had been rather a wag in his younger years who had gradually slumped into an inelegantly debauched middle age, it does not seem that William was very endearing at any point in his life. The public never really took him to their hearts, and they never had much of a chance because by 1837 he too was dead.

By the inexorable laws of male primogeniture, the monarchy came to rest on that lone little daughter of the fourth son, and Alexandrina Victoria became Queen Victoria at the age of eighteen. Victoria had become the single fragile thread linking the British monarchy's past with its future. Told of her place in history as a child she had been shocked, but by the time of her accession in 1837 she was reconciled to her responsibilities. It seems that from the moment she married her charming cousin Albert in 1840, her life and monarchy finally fell into place. The public would not brook a king who was not a sovereign himself, and so her beloved Bertie was to be known as Prince Albert, although he was soon playing the role of the monarch in all but name, as Victoria embarked upon a prolonged period of motherhood.

Between 1840 and 1853 she gave birth to Victoria, Edward, Alice, Alfred, Helena, Louise, Arthur, Leopold, and Beatrice. The single thread perpetuating the monarchy gave rise to many strands, and yet

Victoria's prodigious output of infants was, with hindsight, rather unfortunate. She had inherited her father's star-struck X chromosome and was busily distributing it to her own children—at least three of them, and perhaps as many as five. Yet at the time, the growth of the royal family was seen as a symbol of inward national stability, just as the family of nations coerced into the British Empire was a sign of outward achievement. The people had been unsure at the arrival of the young queen after the demise of her hated uncles, but slowly she edged her way to the core of the British self-image. She was to rule—mentally intact, indeed—until the twentieth century.

Victoria had arrayed a supporting cast of mini-royals around her—Victoria her confidante, Edward her heir, beloved Arthur her favorite, and Beatrice her perpetual baby. She never tried to maintain any pretense of treating her children equally, and they all had to occupy her chosen roles in her life. Victoria was proud of her children, as she had long worried about the vitality of the royal bloodline, and she fully believed that she had done her part to reinvigorate the family stock. One can imagine her reaction when it became clear that all was not well with little Leopold.

One of the later additions to the family, Leopold always had the odds stacked against him—arriving as he did between darling Arthur and Beatrice. Yet just as every toddler learns to fall at the same time as they learn to walk, Leopold also had to learn that falling could mean days of intense pain. The slightest knock could leave him bedridden and sickly for weeks. Victoria's reaction to her weakest child was almost premonitory: "This disease is not in our family!" Little Leopold had hemophilia, which was later to become known as "the royal disease." Victoria's likely reaction to this epithet can hardly be imagined.

Descriptions of a bleeding disease affecting only men appeared occasionally in newspapers and medical journals—the first report was probably published toward the end of the eighteenth century. By the time of Leopold's first fall, doctors were already piecing together the way that the disease runs in families, although they were still erroneously supposing that its root cause was a weakness of the blood vessels. Remarkably, an impressively ancient understanding of hemophilia was apparent from that treasure house of arcane medical wisdom—Jewish law and custom. Late in the Roman era, Jewish writers were advising that any boy should be exempted from circumcision if two of his older brothers had died as a result of bleeding after the procedure. As we will see later, a stipulation that this rule should even apply if the brothers had different fathers is astounding proof that Judaic understanding of disease was to remain unmatched until the twentieth century.

Leopold tried to make the most of his life, but he was thwarted at every turn by his mother. Almost as if she was ashamed of him, Victoria kept him out of the public eye. When he reached adulthood, still he was not released—he was enlisted as Victoria's personal secretary, which mainly involved making copies of her extensive correspondence. To the world outside, it was almost as if poor Leopold did not exist. Although his mother tried to curtail his activities by redirecting his income to his brothers, still Leopold tried to escape, and sometimes he succeeded and briefly tasted the life of any self-respecting nineteenth-century European prince—women, drink, and gambling.

His life took a turn for the better when he married Princess Helen of Waldeck in 1881, and soon a daughter, Alice, was added to the family. Leopold had used his marriage as a route out of subservience to the queen, had become Duke of Albany, and was even tipped for the governorship of the Australian state of Victoria when he died in 1884 at the age of thirty-one. A sign of his new-found freedom, he was gambling in Cannes when he fell down the stairs of a casino. A blow to his head led to uncontrolled bleeding inside his skull and the pressure on his brain quickly killed him. The royal disease had claimed its first victim.

Victoria was to reign until 1901—the year that Clarence McClung was to first suggest that the X chromosome controls the sex of children. Apart from Leopold, all her other children—five daughters and three sons—were apparently healthy. Many of them were to contribute to the royal lines of Europe, marrying into, or bearing children who would marry into, the royal families of Greece, Russia, Germany (repeatedly), and Spain. And of course Edward was, by the rules of male primogeniture, to become Edward VII—a direct ancestor of the present British royals. No doubt most of the family were satisfied for a time that poor Leopold's affliction was just the sad result of random chance—that he was just the runt of the litter. Yet even before the end of Queen Victoria's reign, ominous signs were appearing that Leopold's affliction could jump across the generations.

Almost living up to her mother's prodigious record of fecundity, Victoria's third child, Princess Alice, had six children, yet the old malaise was not to surface until the fourth—a boy named Frederick. In a frightening echo of his uncle, at the age of three Frederick nicked his ear and the wound bled for three days before finally being staunched. Only a few months later he fell from a window and rapidly bled to death. Worse was to come when Frederick's older sister Irene married Prince Henry of Prussia. One of their sons, Henry, clearly suffered from the same old bleeding disease and yet somehow managed to survive into his fifties. His brother Waldemar was not so lucky, and bled to death at the age of four.

Yet it was through another of Alice's children that the "royal disease" was to progress beyond simple family tragedy to alter the course of world history. Alice's fifth child, Alix, or Alexandrina, married the Russian Czar Nicholas II in 1894. The fragile thread of succession that had led to Queen Victoria was now weaving itself deftly into the lives of future sovereigns of Europe. This, of course, was the *raison d'être* of royal families—for hundreds of years they had bound the continent of Europe together, intermarrying to cement alliances, gain influence, or placate enemies. Even in the nineteenth century, the hopes of individual royals were still often subjugated to the strategic exigencies of nations. But by accepting a descendant of Victoria, other royal houses could unwittingly be playing a dangerous game. The Russians played that game and lost.

As soon as Alexandrina and Nicholas were married, the pressure was on for the generation of an heir, preferably a male one. They certainly tried their best, producing Olga, Tatania, Marie, and Anastasia in fairly quick succession. Eventually the much-needed boy appeared—the Czarevich Alexis was born in 1904, Victoria's daughter's daughter's first son. Alexis too was to have hemophilia, but he was not to have the chance to die of it.

Alexis's childhood was miserable. Despite all the protection and care that could be afforded to the single male link to the hoped-for next generation of the Russian royal family, he was often bed-bound with dark bruise-like swellings under his skin and severe crippling pain in his limbs. His blood could not stop itself uselessly bursting into the joints between his bones, causing them to stiffen and ache. The young Czarevich frequently fainted from the intensity of the pain he suffered.

His parents were distraught that their valuable little boy was so terribly afflicted, and the family turned in on itself to try and cope with the heir's suffering. The royal couple almost turned their back on their country in their desperation to see their son cured, and they grew ever more unpopular. The situation became worse when Alexis's parents turned to a monk, one Grigory Rasputin, who claimed to be able to cure the boy by spiritual means. Rasputin gradually developed a hold over the Czar and his wife that was probably a decisive factor in their downfall. The Czar's gradual neglect of the governance of his country was to have ramifications almost beyond belief.

The deep instability and ingrained resentment that had become the basis of Russian political life were cruelly exposed during 1914 to 1918, when the nation was at war with Germany, the Czar's wife's homeland. Almost worse than the alien forces which surged far into the Russian heartland was the sense of helplessness that gripped the nation when those forces were engaged. As the Russian soldiers launched into the slaughter of the war, they knew that behind them their home-towns and cities were starving, and starving mainly because of the inept way in which the war effort, and the country, were being run. Toward the end of the war there were food riots in many Russian cities, and the now-hated Czar was forced to abdicate. In 1918 came the Bolshevik revolution, and any vestige of the old establishment was wiped out. On the sixteenth of July, near Ekaterinburg, the Czar, his wife, their cherished son, and all four daughters were shot. There could be no return to monarchy in the Soviet Union. The thread was broken.

Historians often argue about how world history would have differed had the 1918 revolution never happened, but it is not unreasonable to claim it as the watershed of the twentieth century. And its causes can be directly traced back to Edward Hannover's celestial accident that hot July night one hundred years earlier.

The history of one more country was to turn on the whim of one of Victoria's rogue chromosomes. Like all of Victoria's daughters, baby Beatrice seemed exempt from the disease that was attacking apparently random male members of her family, but she too was another family disaster in the making. Against her mother's wishes, she married Prince Henry of Battenburg in 1885, and they were to produce four children. Once again, all seemed well at first. Their first two children, a boy and a girl, were healthy, but their next two sons,

Leopold and Maurice, also fell victim to the same disease as their uncle Leopold.

Despite her worrying heritage, their daughter Victoria Eugenie was to have an important genetic destiny. Indeed, she was deliberately selected by the Spanish royal hierarchy to infuse (or should that be transfuse?) some fresh blood into the Spanish line. In 1906, Victoria Eugenie survived a wedding-day assassination attempt and married Alfonso XIII. Within ten years of procreation and dodging more assassins they had six children—four boys and two girls—but once again a royal family was to be struck by hemophilia. As usual the girls were spared, but the oldest and youngest boys, Alfonso and Gonzalo, were hemophiliacs.

Spain was in tumult when the princes and princesses were young. Although Alfonso XIII was autocratic, he was not entirely unpopular, partly because Spain successfully distanced itself from the carnage of the First World War. Yet resurgent anarchists, socialists, Basque and Catalan separatists, and a protracted war in Morocco were continual problems. In 1923, he supported a bloodless coup that led to a period of military dictatorship, but when this fell in 1930 and republicans made gains at the ensuing elections, Alfonso XIII was declared a traitor and forced into exile in Italy.

Hemophilia was now to conspire with other family tragedies to mold the Spanish succession into its modern form. Prince Alfonso died in a car crash, leaving the second son Don Jaime next in line for the throne, but he renounced his claim in 1933 because of his deafness. This left Alfonso XIII's third son, the uniquely healthy Don Juan, to carry on the line. In 1938, as the Civil War tore a swathe through the Spanish population, Don Juan's first son was born in Rome. This child was named Juan Carlos, and that baby is now the king of Spain.

At the end of the Civil War, General Francisco Franco came to power. He could not countenance Don Juan's return, but the general and the exiled prince allowed Juan Carlos to return to Spain as a king-in-waiting. Franco was to rule Spain until his death in 1975, but only two days later, Juan Carlos ascended to the throne and embarked on a dramatic program of liberalization of Spanish society. Juan Carlos has not been just a figurehead, and indeed he will probably be most remembered for being instrumental in fending off a military coup in 1981. Yet but for the random destructiveness of his great-great-great-grandfather's chromosome, he would not have become king at all.

Officially, Edward Hannover's chromosomal legacy is now spent—his male descendants either healthy, exsanguinated, or executed. Occasionally rumors appear that suggest that the disease is still trickling unseen through the Spanish royals, but any deaths of obscure infant princes are never ascribed to hemophilia. Perhaps the royal hemophilia nightmare really is over, but for thousands of other families around the world, similar nightmares are only just beginning.

Genes that Jump Generations

How can a disease trickle down the generations of a family in such an insidious way? The spread of Edward's ill-starred X chromosome did not scatter disease randomly throughout his descendants, but instead there was a pattern to its rampage through the royal houses of Europe. The sinister way that hemophilia tracked down some regal lineages while leaving others untouched was both striking and important. Important because, as you read this book, hundreds of other diseases are creeping in exactly the same way through other, less blue-blooded families.

Mothers, fathers, daughters, sons. Hemophilia treats each of them in a very different way, and the reason for this apparent familial unfairness is that the disease is transmitted by a damaged X chromosome. You will remember from the last chapter that the X chromosome is in a unique situation—it is the only chromosome that carries lots of useful genes, but that is also inherited differently by boys and girls—it is, after all, a sex chromosome. The X chromosome is important in hemophilia because it contains two genes that carry the codes to make two proteins which go by the rather uninspired names "factor VIII" and "factor IX."

Every day we all damage many blood vessels as we walk around bumping into things, or even as our knobbly food squeezes its way along our intestines. Coping with bleeding is a continual challenge to our bodies and because of this our blood is in a constant state of readiness to plug up any holes that appear in our circulatory plumbing. Of course, blood must not clot too easily or our healthy blood vessels will get blocked by mistake, which is pretty much what happens when people get thrombosis or heart attacks (remarkably, hemophiliacs are thought to be partially protected from heart disease). Clotting must be controlled very carefully and both the blood and the vessels through which it is pumped contain a large number of pro-clotting and anti-clotting chemicals that keep the whole process in check, triggering clotting only if necessary. Not surprisingly many of the clotting system chemicals are geared to detect damaged blood vessel walls—and factor VIII and factor IX are two such chemicals.

Therein lies the problem of hemophilia. These two vital leak-detecting proteins are made according to the code of two genes carried on the X chromosome, and as we know, that means that while XX girls get two copies of each gene, XY boys only get one. If a baby boy gets a dud factor VIII or factor IX gene, then his blood will not clot properly. If he has an accident, like Prince Leopold, or sometimes even if he ruptures minuscule blood vessels in his joints while simply

walking, like Czarevich Alexis, then the result is often rapid hemor-
rhage. Hemophiliacs do not usually die of their very first bleed, as
other components of the clotting system remain intact, but the ability
to clot is so seriously compromised that injuries can take hours or
even days to stop bleeding. Eventually, many hemophiliacs suffer a
larger injury, and they simply run out of blood before it can clot.

So because they get just one X chromosome, boys are generally
either healthy if the factor VIII and IX genes are intact, or hemophiliac if they are damaged. The situation may be clear cut for boys, but
for girls things start to get more complicated. Girls inherit one X
chromosome from each parent and so it is extremely unlikely that
they will get two damaged X's—there simply are not enough damaged X chromosomes in the population for this to happen very often.
More commonly, a girl will inherit one damaged X chromosome—
she will not be able to make clotting factor from that X, but she will
still have another, normal X that she can use. So unlike boys with a
damaged X chromosome, girls in the same situation do not usually
show the symptoms of hemophilia. (Although careful examination of
the blood of such girls does often show it to be subtly different from
that of girls with fully intact clotting-factor genes.)

This difference between sons and daughters picks out the essential
unfairness of these sex-linked diseases. We saw in the last chapter that
far back in our evolutionary history, X and Y probably started out as
an equal set of non–sex chromosomes carrying a large complement of
genes. Then, however, the unusual specialization of Y caused it to lose
most of its genes, leaving the X to carry them alone. The end result of
these evolutionary shenanigans is that XY boys are now at a tremendous disadvantage. All non–sex chromosomes come in neat pairs, so
that if you inherit some damaged genes on these chromosomes (as
everybody does), then you will have a good copy of each gene to fall
back on. The same is true of women's X chromosomes—girls get a

spare "back-up" copy of every gene on the X chromosome. The only exception to this system is a boy's X chromosome. If part of that is damaged, then the child will simply have to cope the best he can.

So the evolution of chromosomal sex determination has introduced a divisive disease discrepancy between boys and girls, and indeed sex-linked diseases can occur in any mammal that decides its sex in the same way as us. You would not expect sex-linked diseases to occur in the same way in creatures that use egg temperature, for example, to control their sex. Human boys have lost the back-up gene-duplication method of coping with damaged genes on their X chromosome, and now they must live with the consequences—many of our commonest hereditary diseases are sex-linked. There are hundreds of genes on the X chromosome, and hundreds of sex-linked diseases that can occur when those genes get damaged. Our neat way of determining our sex has left us in a terrible predicament. The "macho Y" method of sex determination does not seem like such a good idea now.

Victoria's family also shows us how sex-linked diseases are not only different for sons and daughters, but they are also very different for mothers and fathers too. The same rule that means that men with damaged X-borne genes always show signs of that damage also means that a man in a family afflicted by a sex-linked disease always knows his genetic status. If he does not have the disease, then his X chromosome is normal and he need not worry about passing the disease on to his children. Victoria's first son, later crowned Edward VII, was in exactly this position, and this is why we can be sure that hemophilia was not passed down to the British royal family—indeed, succession by male primogeniture is an excellent way of removing sex-linked diseases from a noble line.

Alternatively, hemophiliac men can be equally certain of their chances of passing on the disease to their children—and this is an

important issue now that they often survive to the age when they can
sire their own children. They cannot transmit the disease to their
sons, as a father always gives his son his Y chromosome, not his X.
This is why Leopold's son Charles was unaffected by the disease.
However, fathers always donate their damaged X to their daughters.
Of course, those daughters will not have the disease, but they can in
turn pass the X chromosome on to their own children.

And this is where the great unknown comes into the equation—
mothers. The same mechanism that stops women getting sex-linked
diseases also means that a woman in an affected family usually does
not know whether she is carrying the damaged gene or not. In the
past, the only way she could find out was to have children, and then
watch her sons closely. This was exactly the predicament of not only
Victoria, but her daughters Alice and Beatrice and her granddaugh-
ters Irene, Alexandrina, and Victoria Eugenie. Perhaps this was also
true of the four young Russian princesses executed at Ekaterinburg,
but we will probably never know if they carried the damaged gene
that made their brother's life such a misery.

So mothers in families affected by a sex-linked disease are in a difficult position. They could have entirely normal X chromosomes, and
so there would be no reason why they should not have children.
Or alternatively they could be carriers of the damaged X—they will
be entirely healthy themselves, but will contain within themselves
the seed of disease in their descendants. Mothers always give their
children one of their X chromosomes at random, and so half of a
carrier woman's sons will have the disease and half of her daughters
will themselves be carriers—in the same uncertain position as their
mother. This is why there were fears that hemophilia could have
persisted in the Spanish royal family. Juan Carlos had hemophiliac
uncles, and so there is no reason why his two aunts could not be
carriers.

This is how sex-linked diseases claw their way through the genera-tions—not through diseased men, but by hitching a ride in appar-ently healthy women. Until recently, the only way that a family could be certain that it had cast off one of these diseases was when, by chance, only healthy male descendants produced children. Today, much of the uncertainty has now been lifted from families affected by sex-linked diseases—women can now often be told with certainty whether they are carriers, and approaches are even being developed to identify embryos with damaged X chromosomes. In fact, the unusual way that X chromosome–borne diseases creep through families may be the most important weapon in our armory to fight them.

I keep saying that sex-linked diseases spread in an unusual fashion, but in what way is it unusual? The X chromosome may carry plenty of genes, but these are still only a small fraction of all the genes we pos-sess. Humans have twenty-two pairs of non–sex chromosomes in every cell, each of which looks very like the X, and it is these that carry most of our genes. Many inherited diseases occur because genes on non–sex chromosomes have become damaged—and these diseases are not sex-linked. Men and women are equally likely to be affected, and these diseases flutter through the human population in a very dif-ferent way to sex-linked diseases. Each of us has two sets of non–sex chromosomes—we get one set of genes from our mother and one set from our father. When a baby inherits damaged copies of the same gene from both parents—a relatively rare event—it is left with no functional copy, and it may suffer from a genetic disease.

Cases of non-sex-linked disease appear sporadically in the popula-tion when a man and a woman who both carry a damaged gene just happen to meet and mate. These parents often show no outward signs of disease because they each have one normal copy of the gene, and the birth of the afflicted child seems to come like a bolt from the blue. However, if that child survives to the age when it wants to have its

own children, then in all probability it will meet a partner with
undamaged genes and all their children will be healthy, although all of
them will then carry a single copy of the damaged gene. These damaged genes will then be dispersed into future generations, where very
rarely they will meet another damaged gene and cause the disease
again.

So whereas sex-linked diseases seem to creep along discrete family
lineages, most other genetic diseases just crop up as rare, apparently
random occurrences. A damaged X chromosome is much more
apparent because approximately half of the boys in an affected family
will show signs of disease, and this continual production of sickly
children can often lead to the family line dying out—thus eradicat-
ing the damaged gene. Because of this, sex-linked diseases are often
relatively short-lived—popping into existence when someone like
Edward Hannover has an unfortunate meeting with a cosmic ray, and
petering out within a few generations. In contrast, damaged non–sex
chromosomes can rattle about in the population for very long periods, usually harmlessly carried by healthy people and only occasionally causing disease.

Yet fortunately, the characteristic way in which sex-linked diseases
spread through families may also lead to their downfall. Unlike other
genetic diseases, it is often very easy to identify people who are at
increased risk of passing on sex-linked diseases to their offspring. The
only people who are likely to do this unknowingly are, of course,
women—if a woman has a sibling or maternal ancestor's sibling with
the disease, then she is clearly one such person. Very often, analysis of
her DNA will give a clear indication of whether she is a carrier or not,
and even when this is not possible, she can at least be told the probability that she is a carrier and the likely effects of this on her children.
In the Epilogue, I discuss how with the help of modern reproductive
technology women carriers can now be helped to selectively conceive

female babies. This can prevent the birth of boys with sex-linked diseases, but it does not stop the birth of carrier girls, and in future it is likely that women will be able to select both male and female embryos that lack the damaged X chromosome altogether. So, the good thing about sex-linked diseases—if there is anything that can be said to be good about them—is that potential carrier individuals are very easy to spot, so that all our technological wizardry can then be unleashed to help them conceive healthy children.

One thing remains uncertain about damaged X chromosomes: where do they come from? Why do sex-linked diseases suddenly flash into existence? You will probably have realized by now that there is actually no concrete evidence that one of Edward Hannover's testicles suffered a head-on collision with a cosmic ray. I chose that as the cause of his descendants' problems partly because it made a good story, but also because it emphasizes the random way in which such mutations occur. In fact, most mutations in the factor VIII and IX genes are probably simple changes to a few rungs of the DNA molecule resulting from some copying or splicing error. But I prefer the celestial to the mundane, and so I am sticking with the supernova story.

We are not even sure that Victoria got her damaged X from Edward. It could have come instead from her mother or one of her maternal ancestors, but this is probably slightly less likely than it coming from her father. Recently, it has even been suggested that it is simply too unlikely that such a rare mutation should appear spontaneously in the sperm destined to help create a future monarch. The implication of this theory is that Victoria was illegitimate, fathered by her mother's hypothetical hemophiliac lover. This would certainly cause some constitutional problems as it would mean that she was ineligible to be queen, and so the entire present British royal family would be a bunch of impostors. My problem with this theory is that although the mutation that damaged the royal X was an unlikely occurrence, is Edward's wife's liaison with a hemophiliac paramour within a month of her marriage any more likely? Unfortunate as it was, that damaged X chromosome must have come from somewhere, and chromosomes often get damaged in testicles.

One striking feature of royal hemophilia is that it was not the result of inbreeding. Diseases that are not sex-linked are often commoner when people in small communities have children with each other over several generations—in contrast, only rarely do sexual partners share damaged genes in a large randomly breeding population. If those partners are closely related, then the chance that they will share such genes is greatly increased, and along with it the chance that their children will be born with non-sex-linked diseases. The royal families of Europe constituted just such a small breeding population, and indeed some of the more unusual characteristics of these families have often been ascribed to an element of regal incest. Yet paradoxically, the most famous royal disease of all, hemophilia, could have befallen even the most gregariously breeding family. Because of the unique way they are inherited, incest is not a common cause of sex-linked disease (although some of the extremely rare cases of hemophilia in girls are thought to result from the union of hemophiliac men and their cousins).

Hemophilia is a very different disease today. It is perhaps one of the best-understood genetic diseases of all, and we now have a treatment for it. Hemophilia can often be treated with injections of factor VIII, but even this apparent godsend for sufferers came at a price. Originally, all factor VIII was prepared from human blood, and every hemophiliac has to inject himself with it very frequently. This meant that for several years, while HIV (the human immunodeficiency virus) spread undetected through the human population, hemophiliacs were routinely injecting themselves with unscreened blood products from many different donors. Because of this, for a large number of hemophiliacs the fear of bleeding to death has been replaced with the fear of developing AIDS. Now that blood can be effectively screened for HIV contamination, such a disaster is unlikely to happen again, and with the advent of artificially produced factor VIII there is no reason why another virus should ever threaten hemophiliacs in the same way.

So men are at risk because they have only one X chromosome, and if that chromosome gets damaged then they have no spare copy with which to make good their loss. Sex-linked diseases move through families in a characteristically devious way, and we now know of hundreds of them. In the rest of this chapter I will explain how men have ended up with an array of X-borne diseases ranging from the almost trivial to the extreme, affecting every part of their bodies. The curse of the lone X is a complex one.

The Vulnerable Giant

There is a giant slumbering on the X chromosome. The largest of its kind, it was infamous long before anyone even knew it was there. Infamous for destroying young men's bodies.

Many children are slow to learn how to walk. Maybe they have other, more pressing plans, like talking, or perhaps they get so good at crawling that they simply do not feel the need to throw off the invisible shackles that tie them to the ground. I was one of those—I was so good at powering around the place on my bottom that I did not deign to walk until I was twenty-two months old. The time that most children choose to walk does not seem to have much effect on their later life—the adult world is not divided into early, skilled walkers and late, incompetent ones.

Yet for one boy in every three or four thousand, this transition to walking is especially difficult, and this difficulty is an ominous sign. Unlike most children who are strong enough to walk long before they actually take to their feet, but lack the co-ordination to so do, these little boys seem to be just a little too weak on their legs. They usually do walk, but the whole process is such an effort that they often push themselves up to standing by pressing their hands against their knees, and sometimes even edging their hands up their thighs.

By the age of three or four, there is usually something obviously wrong. These unlucky little boys have trouble climbing stairs and gradually develop an abnormal gait. It is usually around this age that they are diagnosed for the first time as having Duchenne muscular dystrophy ("dystrophy" means "bad nourishment"). Over the next twenty years their weakness will increase, especially in the top half of their legs, until they are wheelchair-bound. Strangely and characteristically, their calf muscles often swell, even though they are weakening. The creeping weakness also affects their back and it can crumple into a sinuous, collapsed shape. Eventually the degenerating muscles start to scar and tighten irreversibly, rendering large parts of the boys' bodies immobile. The disease also usually affects the muscles of the heart, causing gradual heart failure, as well as the muscles lining the stomach and intestine, leading to digestive problems. Most serious of all, Duchenne muscular dystrophy has the same insidious effects on the breathing muscles and the muscles in the voice box that stop food falling into the lungs. Because of this, many boys with muscular dystrophy die of respiratory infections by the time they are twenty. For some strange reason, the tiny muscles that move the eyeballs are spared, but no one knows why.

For several decades, pathologists have understood what muscular changes are causing this tragic collapse of a young boy's bodily motors. Quite early in the disease, scans and biopsies show that the muscles are attacked by repeated bouts of degeneration that eventually lead to them being replaced by fat and scar tissue. It is this replacement that probably explains the apparently paradoxical way in which some muscles actually get bigger as they weaken. At the same time, the blood is flooded with a protein called creatine kinase, which doctors know is a classic sign of the contents leaking out of damaged muscle. Indeed, the presence of this protein in their blood is often the first clear indication that these little boys are indeed suffering from muscular dystrophy.

Soon after muscular dystrophy was recognized as a distinct disease, it was realized that it occurs in familial clusters. If an affected boy has sisters, then their sons and their daughters' sons are likely to develop the same disease. Yet only very rarely are girls affected—although blood tests sometimes show suspicious increases in creatine kinase. If this all sounds rather familiar, then this is because Duchenne muscular dystrophy is spread in exactly the same way as sex-linked hemophilia—the disease is passed through the generations on a rogue X chromosome, carried silently by double-X females and afflicting poor single-X males picked out by the hand of fate.

Yet unlike hemophilia, which sufferers can sometimes survive long enough to pass on to their daughters (fathers do not usually bequeath their X to their sons), muscular dystrophy is much more of a strictly female-carrier, male-sufferer disease. This is because it is severe, and most of its sufferers are debilitated or dead before they can father children. Duchenne muscular dystrophy is almost never passed from father to daughter—so men really play no part in spreading it, and there is also little scope for the incestuous production of girls with the disease either. The effects of this greater severity are even more noticeable now that effective treatments exist for hemophilia, because muscular dystrophy remains largely untreatable.

Hemophilia is not a unique exception to the usual genetic rules. Duchenne muscular dystrophy is yet another ruinous way in which a damaged X chromosome can make boys' lives a misery. And I am afraid that this has turned out to be a predominant theme of hereditary disease in humans. By taking on its strange role as passive counterpart in controlling our sex, the X has consigned half of us to a catalogue of disease. And sadly, Duchenne muscular dystrophy and hemophilia are not the only often fatal sex-linked diseases.

Yet perhaps because muscular dystrophy is such a depressingly relentless disease and is so relatively common, it has become the ultimate target of modern genetic medicine. The scientific onslaught that has been needed to track down the cause of this disease—to bring our understanding to a level where we can even think about ways of treating it—has been uniquely protracted and ferocious. Sometimes it seems as though if we could conquer Duchenne muscular dystrophy, then we could probably overcome any genetic disease. Everything about it seems to be deliberately designed to confound our attempts to deal with it.

At the start of the attack on Duchenne muscular dystrophy, all scientists knew was that it acts as if it is carried on the X chromosome, and that it affects muscle. But there was a yawning gap of knowledge between these two facts, and that gap would have to be bridged before we would stand any chance of defeating the disease. There was no culprit gene, nothing encoded by that gene, and no comprehension of how that gene product might be involved in the everyday operation of healthy muscle—nothing.

Instead of trying to identify what exactly was wrong with dystrophic muscle, and then working backward, as it were, to the gene, it was realized in the late 1970s that the best way to approach the disease might be to find the gene first. Geneticists crept into nature's laboratory and found a few isolated cases of X chromosome damage that helped them find the gene that is damaged in muscular dystrophy. In particular, they studied rare cases of women who suffered from the disease, apparently because damaged regions had been spliced from one of their X chromosomes onto the other. Both by studying these women and by a process of elimination they gradually narrowed their search until they found the gene—dystrophin.

And what a gene it is. Dystrophin has turned out to be unexpectedly large—huge in fact. You may recall that the DNA molecule on which the genetic code is written is rather like a ladder, and the code is actually carried on the rungs of that ladder. Well, when the rung-code, or sequence, of dystrophin was finally read in the late 1980s, it was found to contain 2.5 million rungs. That is a lot of rungs for any ladder, but even now that we know the sequence of all the human genes, dystrophin is still the largest human gene known—in fact it is the largest gene ever found in any mammal. Dystrophin is the giant of the gene world. It takes up about one sixtieth of the X chromosome, and over a thousandth of all the genetic material in every human cell— that genome we keep hearing so much about in the news. Considering that the genome has to carry maybe thirty thousand genes, as well as a much larger amount of non-gene DNA, dystrophin is clearly taking up a disproportionate amount of space. And for some reason, this behemoth rests on the X chromosome.

Geneticists know the code that cells use to convert genes into proteins, so once they knew the rung-code of dystrophin, they could immediately predict the shape of the protein it encodes. As with most other genes, cells read dystrophin's rungs in sets of three, with each set defining one amino acid building block to be included in the protein chain. So the cell's internal machines (and indeed the geneticists who study them) simply read off these rung-triplets and add the appropriate amino acids sequentially to the growing protein chain. Once the end of the code is reached, the protein is released to do its job within the cell.

As three rungs encode each amino acid, you might predict that the 2.5 million rungs in the dystrophin gene would encode a protein containing about 800,000 amino acids. In fact the actual protein product of dystrophin is very much smaller—some way short of only 4,000 amino acids. The reason for this discrepancy is that only small portions of the huge dystrophin gene region actually contain any meaningful DNA code—it is chopped into seventy-nine coding fragments (exons) separated by much larger regions with no code (introns). In other words, the vast majority of the gene is not used to make anything at all. So most of the time the scientists spent slavishly reading the sequence of dystrophin, they were reading the largely irrelevant separating regions.

This may appear to be an incredibly wasteful way of making a gene—bulking it up with unused noncoding rubbish so that it occupies a much larger amount of space on the chromosome. Yet this seems to be a very popular way of making a gene. In fact, almost all animal genes are punctuated by noncoding regions, although admittedly not usually as many as are found in dystrophin. Geneticists do not know why our genes have these gaps in their protein-making code, but the fact that almost all genes do contain them suggests that the gaps must be there for some reason. Dystrophin simply seems to have lots of gaps, and extremely long gaps. Because of this, dystrophin may be the largest human gene, but it does not encode the largest protein—for example, the product of another gene used in muscle, titin, is much larger. Yet the sheer size of the dystrophin gene does seem rather suspicious. Could it be so vulnerable because it is so large? The larger a gene is, the more there is to get damaged, after all.

Even before the full rung-code of the dystrophin gene had been elucidated, geneticists could tell that the protein it encoded looked very much like the sort of thing that was found in muscle. Of course, this was hardly unexpected, as they had originally tracked down the gene because it was implicated in muscular dystrophy. But if they were ever to fully understand the disease, they had to find out what the dystrophin gene's protein product actually does in healthy muscle.

Muscle is, it turns out, a rather unusual part of the body. Most of the body's muscles are attached to bones, and their job is to move the body around by yanking on strategic bits of skeleton. There are also other equally important muscles that do not move bones—some make up the large muscular pump that is the heart, while others squeeze food along the intestines. Yet all the different kinds of muscle cells really only have one job, and that is to get shorter on command—to contract. Because of this, they have a very specialized structure. Skeletal muscle cells, for example, are long tubes packed with a neatly arranged lattice of contractile machinery made up of, you guessed it, special proteins. When a muscle contracts, some of these proteins actively crawl along others, and it is this crawling that ratchets the ends of the muscle cells together, shortening them. The contractile proteins are anchored at each end to the membrane surrounding each muscle cell, and thence onto another muscle cell, or a tendon, or a bone—whatever the muscle cell is meant to tug on.

Muscle biologists knew about the internal workings of muscle long before geneticists had read the sequence of dystrophin. So as soon as the code was known, the hunt was on to work out what role the normal dystrophin protein might play in healthy muscle. Rather enigmatically, the 3,685 amino acids of the dystrophin protein seem to make a rod one eight-thousandth of a millimeter long, containing some internal hinges. Although various other possibilities were considered, the rod-like nature of the protein led to the suspicion that dystrophin is part of the scaffolding that holds muscle cells together. Dystrophin was unlikely to be part of the contracting machinery itself, as newborn boys with muscular dystrophy can contract their muscles normally. Instead, most theories about dystrophin now suggest that it is involved in keeping muscle cells intact.

Muscle cells are, for obvious reasons, subject to tremendous stresses and strains, yet like all other cells they are lined by a relatively fragile oily coating, the cell membrane. A layer of oils and fats two molecules thick hardly seems likely to survive the continual thrashing about of a muscle cell, and yet that is exactly what it must do. One end of the dystrophin rod looks as if it is designed to attach to this membrane, while the other looks as if it anchors on the internal framework holding the muscle cell together. As muscle cells contract and relax, they are continually changing length, a process that must constantly wrench and tear at the fragile cell membrane. It now seems likely that the job of dystrophin is to hold the membrane safely in place so that it is not shredded by all this violent movement. This theory would go a long way toward explaining why Duchenne muscular dystrophy affects boys in the way it does. To begin with, they can move their bodies in the same way as healthy boys because their muscular contraction engine is normal. But all that movement repeatedly damages their muscle cell membranes because they are not properly secured by dystrophin. Their muscle cells soon become damaged, and instead of healing they are replaced with fat and scar tissue.

So damage to this one titanic gene sets off a chain of events that cripples, debilitates, and eventually kills. Our foray into the exotic world of the machinery of muscle may have taken us a long way from our starting point, but do not forget that muscular dystrophy is at its root a sex-linked disease, caused by damage to a gene on the X chromosome. The case of dystrophin also shows that there is almost no limit to the sorts of genes that the X chromosome can carry.

Dystrophin, the gene that makes the protein that holds muscle cells together, is about as far removed from the world of sex as one can imagine. Yet the X chromosome has these nonsexy genes in abundance, as is clear from the plethora of sex-linked diseases that result when these genes get damaged. Sometimes, all that seems to link the genes on the X chromosome is that men are in the rather perilous position of having only one copy of them. To compound this danger by placing the largest, and thus potentially the most vulnerable, gene on this very chromosome could be considered an act of perverse recklessness by mother nature. Indeed, muscular dystrophy is evidence enough of the catastrophic consequences of such unfortunate gene placement. Yet, we simply do not know why the X chromosome has ended up with the genes that it has. One thing is clear—there certainly does not seem to have been a concerted exodus of especially vital or vulnerable genes from the X. So there they stay, making many men's lives a misery when they get damaged.

Considering the extravagantly complex layout of the dystrophin gene—seventy-nine different coding regions scattered across an enormous tract of the X chromosome—it is perhaps not surprising that the diseases that result when it is damaged are actually far more diverse than we once thought. For example, there is a less common form of muscular dystrophy, the Becker form, which causes pretty much the same symptoms as the Duchenne form, but which progresses much more slowly. Although becoming progressively more disabled throughout their life, men with Becker muscular dystrophy often survive into middle age and beyond. They do seem to produce some dystrophin protein, albeit sparse or abnormal—yet paradoxically their dystrophin genes are often more extensively damaged than those of Duchenne sufferers.

There are other rarer, and even milder, forms—patients with extremely slowly progressing diseases, or with enlargement of their

calf muscles but no other signs. Even now that the rung-code of dystrophin is known, geneticists still have a long way to go before they
can ascribe particular sets of symptoms to particular patterns of gene
damage. One feature of muscular dystrophy that seemed particularly
enigmatic until recently was that approximately one-third of sufferers
also show mild mental retardation. At first it was difficult to explain
why damage to the dystrophin gene should also affect the brain, but
then it was discovered that the seventy-nine coding chunks of the
gene can be respliced to make a slightly different protein that plays a
role in the functioning of nerve cells. In fact, it is now thought that
different types of muscle and nerve cells chop up the gene giant's
products into not just two but several distinct proteins with different
properties suited to those cells' own needs. And presumably, different
forms of gene damage will affect these various proteins in unpredictable ways. The plot thickens.

Of course, the hope is that greater understanding of the dystrophin
gene will improve our chances of managing this awful disease, but
our knowledge is not yet sufficient to help the average person with
muscular dystrophy. Gene therapy, in which undamaged copies of
genes are inserted into the cells of a patient, offers some hope, but
even in this the dystrophin gene seems determined to foil us. Even
with all the noncoding bits removed, the dystrophin sequence is still
very large, and large chunks of DNA are harder to get into human
cells than small chunks. Also, gene therapy is potentially far easier in
cells that are accessible in some way—squirting genes into the lungs
of cystic fibrosis sufferers is a good example—but it is difficult to see
how dystrophin code could be neatly inserted into all the body's
muscle cells, let alone the brain. Instead, all that doctors can offer
most patients is damage limitation: physiotherapy, physical support,
and operations to sever contracted muscles.

Yet there has been one completely fortuitous discovery, which has given us some optimism that this X-borne nemesis can be tackled without recourse to injecting new genes into people. Early in the 1980s a report was published of a patient with two different diseases. He had Duchenne muscular dystrophy, but he also produced too little growth hormone, a hormone made by the pituitary gland, which sits just below the brain. Growth hormone has a wide variety of effects on human metabolism, and one of these is that it is crucial in controlling the growth rate of children—many cases of growth retardation and gigantism are caused by growth hormone deficiency or excess. There is probably no reason why this boy had both diseases—it was a simple coincidence—but random chance resulted in an experiment that doctors would never have been able to justify. Although he had Duchenne muscular dystrophy, the disease was far less severe and rapid than usual. In fact it was almost as if his lack of growth hormone was somehow alleviating the muscular dystrophy—he was still walking at the age of eighteen.

The scientists who originally reported this case were spurred on to recreate this mixture of diseases in an attempt to develop a way to control muscular dystrophy. They wanted a test patient, as well as a comparison patient who would receive no treatment. One problem with muscular dystrophy, however, is that it is a variable disease caused by a very heterogeneous bunch of mutations, and so it is difficult to find two patients with identical dystrophin damage. To overcome this problem the researchers studied two boys with Duchenne muscular dystrophy who shared almost identical genes—a pair of seven-year-old identical twins. One was treated with a drug that suppresses the effects of growth hormone on the body, and the other received placebo tablets. For a whole year the boys, their family, and even their doctors were not told which child was receiving which tablets. By the end of the year, the disease was almost arrested in one brother, but had continued to progress in the other. Finally the treatments were revealed, and the healthier twin proved to be the one who had had his growth hormone suppressed. This dramatic experiment shows how, even if the gene giant cannot be confronted directly, he may still prove vulnerable.

Before we leave the sad world of muscular dystrophy, let us step back and look the strange way it crops up in human populations. Nature and evolution are pretty dispassionate about tragedies. In fact, the drastic effects of muscular dystrophy have a strong influence on how it is travels through the generations. Because almost no male sufferers produce children, X chromosomes with damaged dystrophin must usually be passed on by carrier women. Yet for each carrier woman, only a quarter of her own children will be carriers—one-quarter will be healthy boys, one-quarter affected boys, and one-quarter noncarrier girls. In other words, because it is so severe, this sex-linked disease is transmitted very inefficiently and so it is usually quickly expunged from the population. Muscular dystrophy does not usually rattle down the generations for long—not even as long as hemophilia. Because of the self-destructive nature of the disease, it does not last long in a family, and so many cases of muscular dystrophy turn out to be completely new mutations, not linked to any other known case. In other words, severe X-borne diseases are among the most capricious genetic disorders known—new cases simply appear at random, caused by an entirely novel pattern of genetic damage.

So if muscular dystrophy and its ilk appear sporadically, and are snuffed out in a few generations, what do they contribute to human evolution? Not much, it seems. Unlike non-sex chromosome diseases, they cannot even be claimed to remove genetic rubbish from the human race. They pop up at random, cause some suffering, and then disappear. But this apparent pointlessness raises another question—if putting genes on the X chromosome is such a risky business, then why do we still have an X chromosome at all? We have already seen how the Y chromosome lost many of its genes, and how mole voles lost it in its entirety, so why not the X? Why have we not simply unloaded all the genes on the X chromosome onto other chromosomes and let it fade away? Men could have a Y and women could have, well, nothing. And there would be no more sex-linked disease.

This idea may appear to be quite compelling, but it seems that few animals have found it so. Although shedding the Y seems quite popular, getting rid of the X is not. There must be some counterbalancing force at work that keeps the X topped up with genes, an idea to which I will return in Chapter 3. Some shadowy genetic benefactor is actively supporting the X and the trouble it causes.

Men: Closer to the Apes?

I suppose that it is a purely male rite of passage, although I did not realize it at the time. I and my fellow seven-year-olds had been only too pleased to be removed from our math class, but as far as I recall, we were not too happy to be spending an English September afternoon dressed only in our underpants. My chilly consternation soon changed to simple bemusement when I discovered that one of the main objectives of this strange ritual was for me to be shown some pieces of paper covered with colored dots.

Fear not. The British private school system may be strange, but it is not quite that strange. There was method in their madness. The afternoon had clearly been set aside to ensure that we were developing physically so that we would become sufficiently manly to run an empire, had there still been an empire to run. The colored dots were just the first of a series of examinations of our developing physiques—

most of which justifiably required our state of relative undress, as well as the ability to cough at strategic moments.

Had I been several years older, I would probably have misidentified the dotty images as an album cover from the recent heyday of British psychedelia. However, at this younger, less confused age, I think I realized that the dots were meant to tell the school doctor something about my eyes. "What's this?" he ventured, of the number five delineated in orange dots surrounded by a field of green dots.

"Er, five," I replied, wondering when the real test was going to start. A couple more dotty characters followed, but the real test never really did begin. I could always see the letters or numbers, and even began to wonder if I had completely missed the whole point of the exercise. But then it was all over and the doctor ticked a box, which I hoped was labeled "normal," and we proceeded to the coughing bit. I now know what that test was for too—you may wish to refer to the word "gubernaculum" in the glossary—and I think it probably explained why ten years later the doctor did not raise any genetic concerns when I dated his daughter for a while. Fortunately he never told her about the underpants episode, as far as I am aware.

The spotty patterns are called Isihara charts and, as most men know, they are designed to help doctors detect color blindness. Some female readers may have seen them too, but they are less likely to have been tested with them. Why? Because color blindness is much rarer in women than in men. And I think that, by now, you may be able to guess why.

For centuries, doctors have realized that some men find it much harder to discriminate colors than others. In particular, red and green are a big problem for men in many families. Some men cannot distinguish them at all, whereas others clearly find it very difficult. So-called color blindness is not unknown in women, but it is far rarer—one in twelve men have the condition, as opposed to perhaps one in 250 women. Although color blindness was really just seen as a strange anomaly—hardly the most pressing medical issue in the disease-ridden world—in 1779, a doctor by the name of Michael Lort was inquisitive enough to study the inheritance of color blindness and even reported to the Royal Society how it spreads through families.

I shall not repeat what the estimable Dr. Lort found, because I think you would find it extremely familiar. Most color blindness is caused by altered genes carried on the X chromosome—indeed it was the first medical condition that was ever discovered to be X borne, in 1912. Yes, red-green color blindness is another sex-linked genetic disease, just like hemophilia and muscular dystrophy, but there is much more to it than that. One thing that I hope has become clear to you about those latter two diseases is their sporadic nature. They crop up, cause some suffering, and then fizzle out over the course of a few generations. Color blindness is much more than that. It may seem a trivial disease compared to those callous strikers-down of youth, but unlike them it is woven into the substance of mankind. Color blindness is not an occasional random tragedy—it is part of what we humans are.

Color-blindness testing is, as I described, a rather uninspiring experience if your vision is normal. Who really puts much conscious thought into perceiving colors? Almost all of us see colors, but few of us could give a good explanation of how we do it. We teach our children the names of colors by example, but do we know that everyone sees those colors in the same way? Blue seems cool and red seems hot, but maybe we have just been told that by other people. Perhaps dark green and pale blue would be a horrible color clash if we had not learned from our earliest years that they make the restful vista of trees against sky. Like flavor, color could be a rather subjective thing.

Yet the superficial bits of color perception are pretty well understood. In particular, we know how eyes gather the objective information about hues for our brains to create a subjective idea of color. From the ancient Greeks onward, philosophers speculated on the nature of color, while all the time artists already knew that almost any color can be made from a mixture of three pigments: blue, yellow, and red. During his university vacation, Isaac Newton demonstrated that white light is made up of a spectrum of six colors, although he added a seventh, probably for reasons of obscure biblical numerology—a reassurance for those of you who wondered in embarrassed silence about the differences between blue, indigo, and violet. It was soon demonstrated, however, that white light, and indeed any color, can be made by judicious mixing of three primary colors: blue, green, and red. Only two decades after Lort's study of the inheritance of color blindness, Thomas Young, the English physicist who had first identified the wave nature of light, suggested that our eyes can see color because we have three independent systems for detecting the three primary colors.

Young's insight was remarkably prophetic. All he knew was that blue and green and red can make any color, and that most of us can see these colors. To suggest that we just use blue, green, and red detectors to decode colors was a masterpiece of simple, clear thought. He has turned out to be pretty much correct, although it has taken nearly two hundred years to prove it. At the back of our eyeball, light is focused onto a light-sensitive layer, the retina. There are several different types of cell in the retina, and most of these are involved in the initial processing of visual information. A remarkable amount of image analysis goes on before this information has even left the eye. However, there are four additional types of cell in most people's retinas and these are the actual cells that detect light—the photoreceptors.

All four photoreceptors have the same basic structure. Apart from all the usual stuff that one finds in a cell, about half of these cells is made up of a stack of neatly interleaved membranes—similar in basic composition to the oily membranes that surround muscle cells. The first kind of photoreceptor will not concern us much here but it is, nonetheless, very important. It is called a "rod" because the stack of membranes is roughly cylindrical in shape. Rods respond especially well to very dim light, but they do not give our brains any color information, which is why colors seem to fade in dim light. Also, they are relatively rare in the center of our field of view, which is why you can see a dim star better if you look slightly to one side of it.

The other three types of photoreceptor are called "cones," because their membrane stacks taper to a point. Especially profuse at the center of our field of view, cones let us see bright light, but they also let us discern color. Just to make Thomas Young happy, there are three types of cone intermixed across the retina, and each type responds preferentially to different wavelengths of light. Really, I should call them long-wavelength, medium-wavelength, and short-wavelength cones, but instead I will just call them (slightly incorrectly) red, green, and blue. Cones are not that strict about the colors to which they will respond—green light causes some excitement inside red cones, for example—but they are selective enough for the brain to be able to calculate perceived colors from the signals coming in from the three different cone types.

The actual ability of cones to respond to those different wavelengths can be traced to some very special molecules anchored in their tiny interleaved membranes. Each cone has a particular pigment on the outside of those membranes, and this pigment has two parts. The first is "retinal," a strange little molecule made from vitamin A, which undergoes a subtle alteration when it is hit by a particle of light.

This is why severe vitamin A deficiency causes vision loss. The second molecule is a protein called "opsin" that holds the tiny retinal, responds to its light-induced tremblings, and conveys them into the cone cell, which then relays them to the brain by way of the optic nerve.

Now here comes the clever bit. The retinal component of the visual pigments is the same in all three different types of cone, but each cone has its own peculiar opsin. The red-, green-, and blue-cone opsin proteins are very similar, but they do differ at a few crucial places—just the odd amino-acid building block here and there. These minuscule differences are sufficient to subtly alter the wavelength of light to which the retinal responds. Retinal stuck on blue-cone opsin, for example, is preferentially activated by short-wavelength light. So color vision is mediated entirely by subtle variations in the three opsin proteins present in the three different cone-cell types. And these three similar but subtly different opsins are made, as you would expect, using the code of three similar but subtly different genes.

We saw earlier that most color-blind people have trouble distinguishing red and green, so what gene is damaged in this form of color blindness? Using the sort of probability theory that allows gamblers to predict the roll of two dice, geneticists were able to deduce from the relative commonness of red-green color blindness in men and women that this condition can be caused by alteration to either of two different genes on the X chromosome. And indeed, the genes for both the red-cone opsin and the green-cone opsin were later found hiding on the X. As it happens, it now appears that approximately three-quarters of all cases of color blindness are caused by defective green-cone-opsin genes, and most of the rest are due to defective red-cone-opsin genes.

Although the mathematical details of all this need not worry us here, this deduction has been subsequently confirmed by the discovery of women who have a damaged red-cone-opsin gene on one X chromosome, and a damaged green-cone-opsin gene on the other—they have normal vision because they have single undamaged copies of each gene, but they can pass the different damaged genes on to two different sons—rendering both boys color blind, but in subtly different ways.

But what of blue? Well, this has been much harder to study because blue color blindness is very rare, but the few cases that do crop up do not seem to show the characteristic pattern of inheritance of a sex-linked disease. This was eventually explained when the blue-cone-opsin gene was tracked down to a non-sex chromosome—run-of-the-mill chromosome 7. And, like most damaged genes on non-sex chromosomes, damaged versions of the blue-cone-opsin gene drift around in the population until a male carrier and a female carrier get together and make babies. By the gaming rules of the genetic casino, about one quarter of these babies will get two aberrant blue-cone-opsin genes and have no functional blue cones. An unfortunate, but very rare occurrence.

So the strange way in which color blindness is inherited results from the fact that the red and green genes are on the X chromosome, often in damaged form, whereas the blue is on a non-sex chromosome and is usually intact. But why all this interest in color vision—after all, it is hardly a matter of life and death? Although it might seem unbelievably perverse that traffic lights are red and green, still one does not often hear of people dying of color blindness, let alone deciding not to have children to prevent them inheriting it. Color blindness may cause problems for people interested in careers as pilots, fashion designers, or advertising creatives, but the condition hardly has the morbid cachet of some of the other diseases we have looked at. Yet there are two features of color blindness that set it apart from those sporadic killers—its mildness and its commonness.

Obviously these two features are linked—if color blindness was rapidly fatal, then no doubt we would find that it was a great deal rarer. Yet the condition's mildness and commonness tell us different things about its spread through the human population. First, mildness. Not surprisingly, because color blindness is a fairly benign condition, sufferers do not die young. This means that unlike hemophiliacs or muscular dystrophy sufferers color-blind men are just as likely as anyone else to have children. Because of this, men spread the condition just as much as women do, and so there is little to stop the damaged genes rattling indefinitely down the generations. However, because all is not fair in the world of sex chromosomes, men and women transmit their damaged cone-opsin genes in very different ways. Fathers pass on their damaged X to all their daughters, but none of their sons (because they get their father's Y), whereas carrier women with one damaged X distribute it randomly to half of their children, regardless of sex.

Enough of mildness; what of commonness? One thing about color blindness is certain: if one in twelve men are color-blind, then damaged red- and green-cone-opsin genes are remarkably common. Being color blind is not some rare freak occurrence—it is a mainstream way of life for a surprisingly large proportion of our species. And different versions of genes do not normally get that common without there being some very good reason. Most human genes come in just one basic form, and it seems that if new altered forms do appear by spontaneous mutation, then the sloshing-about of genes as the population interbreeds soon leads to those forms dying out. Even if we did not know what the red- and green-cone-opsin genes were for, we could still surmise that there is a good reason for their existence in several variant forms throughout the human race. In short, we are coming to believe that color blindness is maintained in the human population because it is somehow helpful for it to be there. But what possible function could it serve? To answer that question, we need to travel back in time to find out whence we got our color vision.

First, meet the relatives. We are most nearly related to great apes (like chimps, gorillas, gibbons, and orangutans) and the so-called Old World monkeys (baboons, macaques, and various other comrades). The monkeys and apes in Europe, Africa, and Asia are a discrete little family of related species with our own distinctive characteristics—such as having a pointy nose (humans, of course, are thought to have originated in Africa). Another rather special characteristic is that we have three types of cone, so we all have three-color vision—we are "trichromats." Remarkably, Old World monkeys, apes, and humans are almost the only mammals with three-color vision—and indeed many mammals do not have color vision at all. Although many non-mammals, such as birds and insects, have even better-developed color vision, we are truly blessed among mammals in this respect.

All the simian relatives on that list have essentially the same visual system as us—rods and three types of cone, each with an opsin that preferentially responds to red, green, or blue light. Considering how closely related we all are, it is perhaps not surprising that our opsin genes are also extremely closely related, and we all keep our green- and red-cone opsins on our X chromosomes, so there is scope for sex-linked color blindness in all of us.

If we work our way back a little further along our branch of the mammal family tree, we come to our next closest relatives, the New World monkeys. Spider monkeys, squirrel monkeys, marmosets, tamarins, and their flat-nosed kin are our evolutionary cousins, as it were, so what sort of vision do they have? As it turns out, we now think that almost all of them have two-color vision—they are dichromats. They have a blue-cone-opsin gene on a non-sex chromosome, and a yellowy-cone-opsin on their X. I have called this opsin "yellowy" because it has to do the job of the red- and green-cone-opsin genes present in Old World monkeys, and indeed it is very similar to them.

You may have already noticed something about the two-color vision of New World monkeys—it is effectively the same arrangement that we find in color-blind men. By losing one of their X-borne opsin genes, men revert to the New World monkey situation. I say revert, because it is generally agreed that Old World monkeys evolved their three-color vision as a modification of an ancestral two-color system, whereas New World monkeys simply stuck with the old arrangement. So color-blind men could almost be said to have taken a step back in evolution. Color blindness has made a monkey out of them.

Climbing (or is it swinging?) further along our evolutionary branch, we come to some slightly more distant relatives, the so-called "prosimian" (meaning, rather condescendingly, "pre-monkey") primates. These are those rather cute things like lemurs, lorises, and tarsiers, and they mark a very neat transition between color-sensitive primates and all those other boring, colorless nonprimate mammals. Although it is rather a generalization, prosimians fall into two groups as far as vision is concerned, and these groups are distinguished by their lifestyles. Some prosimians are active during the day—they are "diurnal." The gorgeous ring-tailed lemur springs, almost literally, to mind. These daylight prosimians are thought to have two-color vision rather like New World primates, and it is worth mentioning that the yellowy-cone-opsin gene still clings tenaciously to its home on the X chromosome in these creatures. The other group of prosimians is nocturnal—active at night—and includes all those creatures with such endearingly enormous eyes, like lorises and bush-babies. Nocturnal prosimians have just one-color vision, as they no longer have a yellowy-cone opsin on their X chromosome. All they have are rods for seeing in dim light, and blue cones for seeing in bright light. Because they have no other cones with which to compare their blue-cone information, they may be effectively insensitive to color—monochromats, just like many nonprimates.

This is all very well, but what does it reveal about the preponderance of color blindness in the human population? Well, if you reverse the furry little story I have just told, perhaps you can see how humans got their three-color vision in the first place. Biologists believe that the small mammal from which primates evolved had just one cone-opsin gene, and that was probably not on the X chromosome. At some very early stage of primate evolution, it seems that a copy of this gene was accidentally transferred to the X chromosome. Now these proto-primates had two cone-opsin genes, and by the vagaries of evolution these two genes soon became a little different. The opsins they encoded clung onto their retinal with subtly different fervor, and they started to respond to light of slightly different colors. Time passed and the two genes diverged—they acquired mutations that made them more different. The X-borne gene encoded opsins more and more responsive to long-wavelength light—green, yellow, and red—while the other gene encoded opsins that detected the blue end of the spectrum. This is the stage at which most diurnal prosimians and New World monkeys find themselves—two-color vision.

But in the monkeys that were to found the Old World side of the family, yet another gene duplication event occurred. The cone-opsin gene on the X chromosome was accidentally copied, and the two resulting genes then started to diverge from each other in the same way. One went toward the red extreme of the spectrum, and the other drifted into the middle, into the green. And that is why we, descendents of those monkeys, can distinguish red and green, while most mammals cannot. This copying of the gene into red and green forms is obviously a convenient way of acquiring three-color vision. A single group of New World monkeys, the howler monkeys, have independently done exactly the same thing—among all their dichromat kin, they are the only ones with full-blown three-color vision.

Yet the story was not quite as simple as that. The parting of the blue-cone-opsin gene and the others had been a clean break—they simply parted company and never met again. In contrast, the green and red genes became neighbors, because the duplication event that created them left them lying next to each other on the X chromosome. Now, having two near-identical genes with similar jobs lying next to each other is a very unusual situation. Most genes are scattered fairly randomly around the chromosomes, and when two similar genes are adjacent, it can cause trouble.

The problem arises when chromosomes are taking part in that chromosomal dance that Henking spied all those years ago—swapping chunks with each other before they are packaged into sperm and eggs. When eggs are made in a female's ovary the two X chromosomes swap chunks with each other just like all the other chromosome pairs, and this is where the problem with the red- and green-cone opsins starts. Because these two genes are so very similar, it seems that the cellular machinery doing all this chunk swapping sometimes deliberately fails to distinguish them. Green gets mistaken for red, red gets mistaken for green, some chromosomes lose a gene, others get an extra one, and sometimes strange mixed hybrid genes are spliced together from the red and green genes.

So the red- and green-cone-opsin region of the X chromosome is often a bit of a mess in humans—a string of red genes, green genes, and hybrid genes. To bring a bit of order to this confusion, it appears that only the first two genes in the region are actually used. If one of these two genes is a normal green-cone-opsin gene and the other is a normal red, then vision will be normal. For example, a man with the following array of genes would not be color-blind because his first two cone-opsin genes are a red and a green:

red, green, green, green

whereas conversely these two arrays would lead to complete red-green color blindness:

red, red, green, green

green, green, green, green

The strange spliced-together red-green hybrids can cause rather more vague forms of color blindness. For example, the effects of the array

red, hybrid, green, green, green

rather depend on the exact rung code of the hybrid gene. If it is more like a green gene than a red gene, then the man may be able to discern colors almost normally. If, however, the hybrid is more like a red gene, then he may be almost as color blind as if he did not have that gene at all.

This is why red-green color blindness can be either total or partial, depending on the first two genes on that X chromosome list. Actually, it is thought that there may be even more color-blind people than we realize, as many cases of partial color blindness may be so mild that they are never diagnosed. Certainly there seem to be more abnormal sets of cone-opsin genes in the male population than there are men who think they are color blind. And all this confusion has resulted because dividing human cells cannot cope with the almost identical red- and green-cone-opsin genes sitting right next to each other on the X chromosome. Like two colors plucked from the palette and hastily applied to the adjacent parts of the canvas, the red and green have a habit of smudging into each other.

Before we find out why red-green color blindness has been built into the human race, it is worth emphasizing just how common it is compared to other forms of color blindness. Like any part of the body, just about anything that can go wrong with color vision does go wrong, resulting in many different forms of color blindness. Yet compared to red-green color blindness all these other conditions are extremely rare. We have already seen that blue color blindness is very uncommon, and the same is true of most of the other possibilities. Sometimes, there is a complete failure of the red-green region of the X chromosome, so that people (usually men) are left with just rods and blue cones—a bit like nocturnal prosimian primates. Other people have no functional cones at all, and so they are extremely sensitive to bright light because all they have to see with are rods, and they are designed for dim light. This no-cone disorder is called "achromatopsia" and it is vanishingly rare in the general population. However, the condition is prevalent on a small island in the Pacific called Pingelap. It is thought that a violent tropical storm laid the island waste in the eighteenth century, leaving only one male survivor, and the island was subsequently repopulated with the progeny of this one man. Unfortunately, he carried a damaged version of a gene normally used to make a protein vital to the functioning of cone cells, so modern Pingelap is full of completely color-blind people.

If we want to know why so many men are born color blind, we need to look at whether our relatives get color blindness as often as we do. I have already described how our closest relatives, Old World monkeys and apes, have a visual system almost identical to ours, and so could potentially be prone to red-green color blindness too. Yet when geneticists actually looked at the cone-opsin genes that these monkeys carry on their X chromosome, they found that they are much tidier than those on the human X. There is certainly not the same inept shuffling and splicing together of the red and green genes that occurs in humans, and in fact as few as one in every thousand male monkeys appears to be color blind. This is a remarkable finding as it shows that whereas normal human vision is pretty much the same as monkey and ape vision, only humans have played fast and loose with the red and green bits. Why should this be? Our ancestors were obviously blessed with reliable three-color vision, so why did we turn it into such a shambles?

Yet we are not the only primates whose males have poorer color vision than females. If we look slightly further afield we can indeed find primate species in which some females have three-color vision, but males have two-color vision. I know I said that New World monkeys (except the howler) have two-color vision, but I am afraid that I fibbed. I can justify that lie by pointing out that, until recently, it was exactly what we believed. These flat-nosed monkeys certainly do have only two places to put cone-opsin genes—the blue one is put on a non-sex chromosome (as it is in humans) and the other, yellowy one is on the X (where our red and green would be). Yet it was recently discovered that female squirrel monkeys can have subtly different yellowy-cone-opsin genes at equivalent positions on each of their two X chromosomes—a red-responsive gene on one X and a green-responsive gene on the other, no less. So, because they have two X's, some females can distinguish red and green, whereas male squirrel monkeys never can. This sexist system is not the same as that operating in humans, but it does show how the positioning of these genes on the X means that it is often male primates that end up having poorer color vision.

If we want to find out why it is apparently acceptable for male primates to lose out when it comes to color vision, perhaps we should try and find out what we actually use our three-color vision for. What do you think of when you think of monkeys? Yes: bananas. Everyone knows that monkeys like fruit, and indeed for many of them it is their staple diet. Indeed, primates like eating fruit more than just about any other mammal, so there is a longstanding theory that three-color vision evolved to help us know when fruit is ripe. Certainly, the somewhat humorous implications of a stomach full of unripe fruit seems like a pretty good reason to develop good color vision. Yet behavioral studies of several species of primates show that they do not actually use red-green discrimination to discern the ripeness of fruit—they apparently do that with their blue-yellow discrimination, which presumably explains why dichromat New World monkeys do not suffer from chronic unripe fruit–induced diarrhea.

Instead it seems that our primate ancestors evolved three-color vision to find fresh young leaves. Many primates eat leaves, but unlike many other mammalian folivores (one of my favorite words—"leaf-eater"), we do not have a huge fermenting gut to digest them. To get over this, many primates are quite selective about what they eat, preferring only the freshest leaves, which contain the most protein and are the least tough. In the tropics, young leaves often flush red before turning green. The tropical forests are big places, and there are not really that many monkeys around, so there does not seem to have been much pressure for plants to lose this red tinge even though it spells "eat me" to any passing whooping trichromat. If three-color vision is mainly useful for selecting leaves for consumption, then perhaps we have one reason why human three-color vision was allowed to degenerate. When did you last see someone eat a leaf? Maybe a reduced dependence on leaves started the gradual deterioration of our color vision.

Or maybe it was more than that. Perhaps there was even a reason for color blindness—an advantage. Although less color information reaches the brain of a color-blind man, it now seems that this can somehow allow him to see more clearly. For some time now, vision experts have discussed anecdotal reports of one alleged advantage of color blindness—that it can allow men to penetrate green camouflage to see moving objects. I must emphasize that this is far from proven, but it has been reported that when groups of men go hunting in the forest it is often the color-blind member of the party who is the first to see the quarry through the undergrowth. It has even been suggested that it is best to send color-blind soldiers out on reconnaissance missions because they are less easily fooled by artificial camouflage—an interesting suggestion, as color blindness is sufficient grounds to prevent entry to many of the world's armies.

Perhaps this is where the male-female difference comes to the fore. Can we find a good reason why these abilities are more useful for men? The problem with answering this question is that no one really knows what men and women were doing for most of the time that our species was evolving. However, it is still a popular idea that we were hunter-gatherers and that men did most of the hunting and women did most of the gathering. Perhaps that makes more sense when one sex is pregnant or breastfeeding much of the time. Maybe natural selection drove one in twelve men to revert to two-color vision to give them the hunting edge, in the knowledge that their female partner would still be able to find, select, and sort the vegetable component of their diet for them. Of course, the aberrant cone-opsin genes could not become too common, or many women would be born color blind too.

I know what you are thinking, and of course it is true. This could be a horrendous oversimplification of what actually happened in human history, and indeed the history of all our primate allies. It could even be rather chauvinistic. Yet there still remain the central undeniable facts of the matter—the red- and green-cone-opsin genes have clung tenaciously to the primate X chromosome, and because of this humans have been able to evolve a selective deficit in color vision in one sex—men. We obviously did not have to do this—after all, none of our closest ape relatives did it. But men and women are undeniably different, and somewhere along the way those differences changed the way we see the world.

That is what the X chromosome has done to men. They only get one X, so they get one copy of each of its genes, so all hell breaks loose when one of those genes is damaged. Unlike their mothers and sisters and partners and daughters, men have no spare copy to fall back on. Although the effects of sex-linked diseases can range from fatal to possibly even useful, it cannot be denied that a single X has left men dangerously exposed, and no one even knows why this ridiculous situation is allowed to continue.

Taking a broad view, men really are the ones who suffer the most because of the X chromosome. Awful though this suffering can be, at least many people now have a basic understanding of sex-linked diseases. Yet the X chromosome does something far more weird to women, and I am sure that this will be news to most of you.

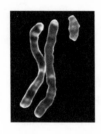

INTERLUDE:

HOW SEXY IS X?

Although you are two-thirds of the way through this book, I can hardly claim to have been comprehensive. A quick glance at the figure below will show you that of the 1,200 or so genes thought to be present on the X chromosome, I have mentioned exactly seven. I hope I can justify this selectivity by claiming to have chosen some of the more interesting of those 1,200. After all, you did not pick up this book to read a list of genes, did you?

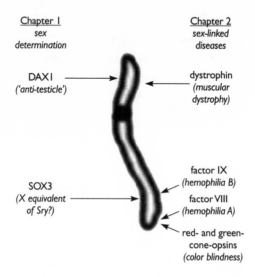

Chapter 1
sex
determination

Chapter 2
sex-linked
diseases

DAX1
('anti-testicle')

dystrophin
(muscular
dystrophy)

factor IX
(hemophilia B)

SOX3
(X equivalent
of Sry?)

factor VIII
(hemophilia A)

red- and green-
cone-opsins
(color blindness)

One thing that may have disappointed you so far is that those seven genes are not, on the whole, very sexy. To be honest, apart from *Dax1* they do not really have anything to do with sex at all— they seem to have had little to do with which sex an embryo becomes, or the formation of characteristically male or female bodies, or the development of adult sexual preferences. But does this really reflect the true nature of the X chromosome? After all, it is a "sex chromosome," so it would be rather disappointing if it did not have anything to do with sex. Of course the macho little Y chromosome has *Sry*, and you cannot get much more sex-obsessed than that, but apart from *Sry* I have not yet discussed any genes related to reproduction on the Y either.

Before we look at what sorts of genes might be found on the X chromosome, it is worth mentioning an assumption that geneticists often make when they start to study questions like these—they assume that genes are scattered randomly among the human chromosomes. With the obvious exception of *Sry*, the actual location of most genes does not seem to be very important. If, for example, ten genes are needed to operate one of the body's internal processes, then they will almost certainly not be near each other, or even on the same chromosome. Close proximity does not seem to be important for genes to work together. There are exceptions to this rule—you may recall that the red-and green-cone-opsin genes are right next to each other on the X chromosome—but in general the rule holds up pretty well. Chromosomal location is not thought to be very important for most genes.

So there is no particular reason to expect the X chromosome to be brimful of sex-related genes. This, of course, made it all the more remarkable when the X was found to be unusually blessed with reproduction genes, especially when these genes turned out to be especially involved in reproduction in men. One research group

looked for genes important in the early stages of sperm production—a fairly specialized process unlike any other in the body. They
isolated twenty-five gene products from immature sperm, and then
proceeded to hunt out where the genes were located among the
chromosomes. The X chromosome contains about one-twentieth of
all the DNA in the body, so, of the twenty-five genes they were looking for, you might have predicted that one or two would be on the X
chromosome. To their surprise, an impressive ten of these sperm-
related genes turned out to be X borne—far more than would be
predicted if they were scattered randomly.

It really does seem that the X chromosome is packed with
reproduction-related genes. In fact, other studies have found more
genes involved in sex on the X, including genes important in female
sexuality as well. Also, genes present on the X have been implicated
in prostate cancer—the commonest cancer in men in many countries. It has even been claimed that a randomly selected chunk of
X chromosome is three times more likely to include a sexy gene
than a random chunk of a non-sex chromosome. Remarkably, sperm-
related genes are so abundant on the X that it is now thought likely
that some forms of male infertility may be caused by damaged X
chromosomes—and so be inherited like sex-linked diseases. And in
fact the Y chromosome has also turned out to be relatively overbur-
dened with reproduction-related genes. So X and Y are seething with
latent sexuality. They really are, after all, sex chromosomes.

Rather satisfyingly, in 1984, long before it was possible to catalogue the contents of chromosomes in this way, a Californian scientist named William Rice predicted all this. In a remarkably prescient
thought experiment, he pondered hypothetical rare genes that are
beneficial to males but not females. He argued that one of these male-
benefiting genes is more likely to spread through a population if it is
present on the X chromosome. After all, being on the X, its presence

would be evident in every man who carries it, and so its helpful effects would promote the success of those men, and thus its own chances of being passed on through the generations. Although X chromosomes spend much of their time in women, any adverse effects of these rare male-benefiting genes in women would almost never become apparent, because most women would have at least one X without the gene. So being on the X improves a rare male-benefiting gene's chances of becoming more common.

Are you with me so far? Rice then went on to show how these genes could become even more common, and even more dedicated to helping men. When such a hypothetical gene becomes reasonably common, then more and more women will be born who have two copies of it—one on each X chromosome. By now, the gene has become so helpful to their sons that the pressure is on for the gene to lose any propensity to adversely affect women. So gradually the gene evolves so that it is simply never switched on in women—it is only ever used in men. This is not such an unusual thing for a gene to do, as many genes are only used by one sex—genes for making milk, or making sperm, for example. Now that the gene does no harm to women, there is nothing to stop it spreading throughout the population. And what genes, we may ask, are only ever used in men? Well, of course, making sperm is an exclusively male preserve, as is building and operating a prostate gland. And those are exactly the sorts of genes that now appear to be proving Rice right a decade and a half later.

To emphasize just how special the X chromosome is, it is worth pointing out that it is hard to see how male-benefiting genes on non-sex chromosomes could prosper if they had adverse effects on females. Men and women have two of each non-sex chromosome, so there is none of that chromosomal inequality that gave those X-borne genes such an advantage. Indeed, the X chromosome almost seems to be actively attracting sexy genes, and in fact the same is true for the Y chromosome. Although we saw in Chapter 1 that the Y has a tendency to lose genes, there are also good reasons for male-benefiting genes to relocate there—not least that they will only ever be used in males, and will never be a burden to females.

So the X is a hotbed of sex-related genes. Yet it also appears to have an unfair complement of some other types of genes too, although the reasons for this are less clear. For example, it has an unusually large number of genes used in muscle cells—dystrophin springs to mind—but it is rather difficult to see why this might be so. As it happens, muscle-related genes seem especially prone to congregating on just three chromosomes—17, 19, and X—so perhaps it is simply a coincidence that X was one of the chromosomes they decided to cluster on. After all, it is as good a chromosome as any.

One especially controversial part of the "How sexy is X?" question has been whether male homosexuality is carried by a gene on the X chromosome. Homosexuality is a fascinating topic for geneticists as it represents an unusual combination—male homosexuals are extremely common, but they are relatively unlikely to have children. We do not know if homosexuality is genetic, but if it were, it would be hard to see why it does not disappear when so many homosexuals do not pass on their genes.

The genetic cat was first set among the moral pigeons with the publication of a study of the inheritance of a propensity for male homosexuality in 1993. A group at the National Cancer Institute decided to investigate an anecdotal suggestion that male homosexuality could be passed down families in a pattern similar to that seen in sex-linked diseases. They studied the families of 114 homosexual men, and found that their maternal uncles and male cousins were more likely to be homosexual than the general population, whereas their fathers and paternal uncles and male cousins were not. This, as we have seen, is the characteristic pattern of a trait carried on the X chromosome. The researchers went even further and claimed that a particular region of the X chromosome—"Xq28"—had distinctive features in homosexual men. They had not found a "gay gene" as such, but they were not far off—or so they thought.

The most important implication of this study was that male homosexuality might be controlled by our genes. This was considered very important by many gay rights activists, and a major blow by many conservatives. After all, genetic homosexuality would mean that men are simply "born gay"—they do not "become" gay during their formative years. This confirmed what many homosexuals already believe, that their sexual preferences were built into them in a very profound way, and were not the result of some alleged developmental aberration. Also, genetic homosexuality would render irrelevant the calls of conservatives to prevent boys being exposed to "homosexual influences" such as gay role models. Why bother, if they are programmed to be homosexual or heterosexual by the time they are born?

There are, however, problems with the Xq28 story. When the study was first published, the location of the "gay-gene" on the X chromosome was claimed to explain why it does not die out. Yet we saw in the last chapter that the fact that a gene can be silently carried by women is simply not enough to ensure its survival. Let us make a rather unusual assumption—that being homosexual has a similar effect on a man's chances of having children as, say, hemophilia. We have already seen that women can carry the gene for hemophilia without getting the disease themselves, yet the fact that many of their male offspring will get the disease is enough to ensure that the damaged gene is usually lost after a few generations. Homosexuality is not, of course, a disease, but there is no reason why it should be any more successful at persisting in a family than hemophilia. So simply being carried on the X chromosome would not explain why male homosexuality is so common.

Although many scientists still believe that there may well be a genetic component to homosexuality, the evidence that it is carried on the X, and certainly at Xq28, is starting to look a bit shaky. To cut a long story short, no other group has managed to repeat the finding that this little chunk of the X chromosome is related in any way to homosexuality. This does not, however, mean that the original study is meaningless. Perhaps those results were a reflection of the men selected for the study. Maybe there are many different causes of homosexuality, and by studying apparent cases of "familial" homosexuality, the researchers were unwittingly selecting just a few of these causes. I predict that in twenty years' time, studies like these will have told us exactly why men are born, or become, homosexual.

Rather strikingly, the quest for the genetic basis of homosexuality has become something of a tussle between the X and Y chromosomes. A more recent, theoretical study has suggested that homosexuality may result from an interaction between a male baby's Y chromosome and his mother's immune system. It has been known for a few decades that one of the genes on the Y carries the code for a protein called H-Y, and H-Y is often recognized as foreign by women's immune systems. Because of this, the authors of this study proposed that as a woman has more and more sons, she gradually develops antibodies against the H-Y that has leaked into her blood from all her male fetuses. Eventually she makes enough of this antibody for it to leach into subsequent male babies and stick to the H-Y in their developing brain. So, they argued, boys with older brothers are likely to have their male brain development altered by their mother's antibodies. This idea may seem far-fetched, and indeed only some of the links in this chain actually have any evidence to support them. However, it does explain one feature of male homosexuality that has been claimed in the past, that it is commoner in boys with several older brothers.

Most of us want to know how much we are a product of our genes, and how much we are a product of our environment, and everyone is interested in sex. Because of its unequal role in men and women, the X chromosome is sure to lie at the center of our genesis as sexual beings.

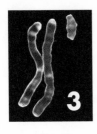

3 THE DOUBLE LIFE
OF WOMEN

Here is a story to confirm some suspicions.

There can be few of us who can claim never to have gasped in exasperation at the unreasonable behavior of a member of the opposite sex. Every day, workplaces, meeting places, and homes resound to the same old stories. Men seem to take a positive pleasure in recounting the unpredictable, capricious nature of women. And, with equal alacrity, women marvel at the unwitting predictability of men. It is almost as if the sexes' mentalities were deliberately designed to irritate and enchant each other in equal measure.

From this point on, I shall no longer attempt to discuss this issue even-handedly. I am, after all, a man, so why should I? Women often seem, to be honest, unnecessarily complex to men. This is not to say that women do not have an equally valid gripe to the opposite effect about men, but obviously I have no firsthand experience of that. There is no point in being fair anyway, as that would go against the grain. The idea that women are somehow more complex than men is deeply embedded in our cultural heritage, and until recently that culture was dominated by the musings of men.

Hinduism is perhaps the most ancient coherent body of philosophy still in existence, and it contains more than a glimmer of the idea that women are inherently mixed beings. Although it includes elements of even-handed male-female duality and equivalence found in mythologies around the world, the great religion of the subcontinent also introduced a marked asymmetry of the sexes into all aspects of life. Perhaps starting with the premise that a man is the basic unit of humanity, women are placed slightly to one side, in a somewhat indeterminate position. Women can, of course, reach Nirvana just like men, but this does not put them on an equal footing. One need look no further than the spiritual tourist's favorite, the Bhagavad Gita, to find an ingrained belief that women are not free, but must be constantly protected by men. It does not say that they should be treated like slaves, but more that they can be considered akin to children, requiring benevolent supervision. Yet they are not slaves, nor children, nor men—they represent some intermediate hybrid state.

At times, the attitude of the European philosophical establishment toward women has been considerably less benign. Again and again, from the ancient Greeks onward, men have been claimed to differ from the animals because they have the power of rational thought, and women have been deemed to lie on the animal side of that divide. By this subterfuge, men ended up nearer to God, and women nearer to evil. Yet women were clearly human like men, despite their claimed animal nature. So where did that leave the poor confused creatures? They seem to have emerged as some sort of mixture of human godliness and animal baseness.

The mythology that appears most comprehensively to have tied itself in knots about the mixed nature of women is Christianity. In fact, a suitably forewarned eye cast over the Bible might come away with the suspicion that this issue was an obsession that plagued the writing of the good book. The two most prominent women in the Bible are Eve and Mary, and they pick out the sexual confusion bubbling inside the book's many authors far better than any psychoanalyst could. From the moment she tasted the apple, Eve became not only the cause of the Fall from grace, but the archetypal female sexual predator, ensnaring innocent, godly men. In fact, except for the rather beautiful erotic interlude of the Song of Songs, the view of women put forward in Genesis is not really redressed until the appearance of Mary in the Gospels. On a superficial level, she is of course a New Testament symbol of virtuous innocence—a counterbalance to the Old Testament Eve. Yet even Mary is not an uncomplicated being—indeed she is an inherently paradoxical mixture—and because of this she has become the obsession of some strands of Christian thinking. She is both virgin and mother, both spouse of God and spouse of Joseph, and even her godliness is confusingly uncertain. As the chosen vessel for the birth of the Christ, she herself attained a uniquely mixed status, somewhere between man and the angels. Believe me, in the centuries immediately after the virgin birth, nations actually went to war over disagreements concerning the spiritual nature of the Virgin and child. The age-old debate about Mary's status even has a parallel in modern studies of the historicity of the Bible—fascinated as they are with the possibility that Jesus had earthly half-siblings: was Mary sullied by fruitful intercourse with her earthly spouse?

Perhaps it is not too surprising, then, that the first scientific approaches to the mixed nature of women were made by scientists immersed in this Judeo-Christian tradition. I have already mentioned, in Chapter 1, the seventeenth-century physician-genius William Harvey, who expressed some strongly held views about the status of women, not too far from the biblical idea—too complex for their own good. Yet, as the father of modern biology, he was also the first to attempt to explain the internal contradictions of femaleness. When he published his *Generation of Animals* in 1651, no one knew how women suddenly "become" pregnant, but Harvey was adamant that this is exactly the time at which their complexity is most apparent. Before they conceive, women are simple selfish sexual beings, just like men. Yet he was fascinated by the fact that, at some mysterious point in time, they switch abruptly to a nurturing, pregnant state, "in the same way as iron touched by the magnet."

That women's bodies appear to have two modes—sexual and nurturing—whereas men's only have one, was hardly a new discovery. But Harvey's attempts to define these modes gave the idea of female complexity a kind of secular, scientific credibility. The biological and sociological study of how women balance these two roles continues to the present day, but at least it all makes some sort of sense. However, the advent of twentieth-century genetics has unearthed yet another dualism that is woven into femininity, and this dualism is both incredibly profound and unspeakably arbitrary. As I hope to show you in this chapter, women are mixed creatures and men are not, in a far deeper way than Harvey could ever have imagined. One of the sexes had to lead a problematical double life, and evolution decreed that it should be women.

A Passable Mosaic

After Henking's chance discovery of the "wallflower" chromosome that has become the subject of this book, we should no longer be surprised when some apparently dry microscopical endeavor accidentally unearths something magical. Nor should we be surprised if the full importance of the discovery is not fully understood for many years.

Just such a discovery was made by the Canadian neurobiologist Murray Barr in 1949. As with many of science's great breakthroughs, this one was accidental. Barr was interested in the effects of fatigue on the human brain, and especially the adverse effects that he suspected might occur in overworked airmen. Barr was studying thin slices of cat brain tissue stained with a chemical called Feulgen, which picks out DNA within the nuclei of cells. Although it was not really important to the job at hand, Barr could not help noticing that some cells contained a little bundle of darkly stained material within their nuclei. At first sight this may not seem very interesting, as cells hold most of their DNA in their nuclei, but something certainly struck Barr as strange. The cells he was studying were not dividing, so most of their DNA was spread diffusely throughout their nuclei—most, that is, except the neat little bundles of DNA that Barr had noticed.

Again and again they appeared, in many of the cells on Barr's microscope slide—each an apparently drumstick-shaped blob of DNA nestling against the edge of the nucleus. Less-inquisitive souls would probably have dismissed these flecks as some strange idiosyncrasy of the feline brain, but not Barr. He searched through similar slides from other animals and he noticed something interesting: some animals had the extra DNA bundles and others did not. And then, crucially, he realized that the bundles were present in female cats, but not males. Barr had made a remarkable discovery—a clearly visible difference between male and female cells. In fact, his drumstick of DNA has turned out to be a consistent feature distinguishing almost all female mammalian cells from male ones.

Murray Barr called the female DNA bundle the "sex chromatin," and he clearly felt that this name was portentous enough. Yet history recalls it as the "Barr body." It is thought that Barr himself was too modest to use this self-aggrandizing term, which is surprising for someone who rather approved of the use of eponymous terms in anatomy. I teach brain anatomy, and so I too have mentally navigated the aqueduct of Sylvius, become dizzy on the circle of Willis, and have even wandered as far as the zonules of Zinn. The brain is rich with the wonderful names of its early explorers, and anyone studying it cannot fail to be captivated by their poetry. And so I will not speak of the undeniably correct but unspeakably dull "sex chromatin," but the enigmatic "Barr body."

This fragment of DNA appeared to mark a cell as female. Women's cells do not contain any special sort of DNA not found in male cells, so what could the Barr body possibly be? Had Barr found his body in males and not females, then of course the Y chromosome would be a good candidate for the strange cellular tenant, but there it was— resolutely restricted to females. During the 1950s geneticists started to investigate the Barr body further, and found that it was not a unique feature of cats' brains. They found it in other tissues and other species—even marsupials—and showed that it first appears when mammalian embryos are no more than a hollow ball of cells containing an as-yet formless embryo. Yet all this still did not give any clue to the true identity or significance of the Barr body.

In many ways 1960 was the best of times and the worst of times for the Barr body. Following fears that men were entering events at the Olympic games masquerading as women, the International Olympic Committee introduced a program of blanket "sex-testing" of all entrants for women's events. At the Rome games this testing involved actual physical examination of the competitors' genitals, and the resentment this caused eventually led to a change in policy. In Mexico City in 1968, the presence of a Barr body in athletes' cells was used as a criteria of femaleness. As I will explain, this indicated a very simplistic view of human sexuality on the part of the IOC and, as a result, the Barr body became more hated than one would have thought possible for a humble pellet of DNA.

On a more positive note, 1960 also brought the first correct identification of the Barr body. After exhaustive study of different cells from female mice, the geneticist Susumu Ohno, working at the City of Hope Cancer Center in Los Angeles, claimed that the Barr body was, in fact, an X chromosome. And Ohno's discovery clearly set some mental alarm bells ringing on the other side of the Atlantic. The very next year, Mary Lyon, a mouse geneticist working in Harwell, England published one of those wonderful papers that, in a short slab of text, propounds a theory linking several disparate and previously mysterious facts. In eight business-like paragraphs she explained why the Barr body is there, and how its presence may have untold ramifications for all mammals, including humans.

First of all, it was clear that mice do not need two active X chromosomes to function properly—male mice are XY, of course, and XO female mice appear quite normal too. Second, Lyon pointed out how Barr bodies suddenly start to appear at a fairly early stage of embryonic life—implying that female embryos do not seem to need them before this time. Her final strand of evidence was probably her most inspired. She pointed out that there are several mousy varieties of genetically inherited fur color which exhibit an intriguing pattern. These color mutations are inherited in a sex-linked fashion because the genes that cause them are carried on the X chromosome. More important, they affect females in a very strange way. Female mice that have a mutant color gene on just one of their X chromosomes show a characteristically patchy effect—half of their fur is the mutant color and the other half is normal, and the two colors occur in random flecks all over the mouse's body.

From this unlikely starting point, Lyon developed her remarkable theory. At a certain point in a female mouse embryo's development, each of its cells "switches off" one randomly selected X chromosome and packages it into a Barr body. From that point on, all the cells derived from that cell will inherit both X chromosomes, but one will be kept as an inactive Barr body, and only the other X will be used for running the cell. As a result, an adult female mouse is made up of two distinct cell populations—each using a different X chromosome to make their coat color. Each dash of color in the coat of a mutant female mouse represents the cellular descendants of a single embryonic cell that chose to use a particular X chromosome. She is like a mosaic of furry tiles of two different colors, and indeed "mosaic" is the rather evocative jargon term for animals made up of a mixture of two different types of cell.

Mary Lyon's most challenging idea was that mosaicism may be found far more widely than just in mice, as X chromosome Barr bodies had also been discovered in rats and opossums, and female-restricted mottled coat patterns were known in other species as well. She chose the calico cat as her example, and subsequent research has shown that she was entirely justified in her choice. In cats, the gene for marmalade fur is carried on the X chromosome, so XY tom cats either have the gene and are marmalade (or what we in Britain call "ginger"), or do not have it and are not. But for queen cats there are three possibilities—they may have two copies of the gene and be marmalade, or not have the gene at all, in which case they will be a different color. However, queen cats with the marmalade gene on just one X chromosome have a mottled coat—with marmalade patches mixed with nonmarmalade patches. This is why we have calico cats, and why they are almost always female. Calico cats are the demonstration of female mosaicism that curls up in your laundry basket.

Lyon was right, and scientists now think that almost all female mammals are X chromosome mosaics. And that means women, too. Almost every woman is, inside and out, a patchwork of two different cells—some using one X chromosome, and some the other. Early in an embryonic girl's existence, each of her cells commits to using just one X and then sticks to that decision, so that each one of the family of cells to which it gives rise uses precisely that X. This does not mean that we get calico women, but women experience far more important effects of mosaicism than any mottled external appearance. An XX woman is a mixture of two different sets of cells using different X chromosomes, as though she were two genetically different animals stirred together.

Only the X chromosome can do this, because no other chromosomes get switched off in an embryo. What more all-encompassing way could one want for women to be more complex than men? All of a man's cells are just about genetically identical to each other, but a woman is a double creature. And the double life of women has far-reaching effects on their health, sexuality, and behavior.

Getting the Dose Right

You may be wondering why this strange state of affairs has come about. At first sight, there may not seem to be any good reason why a woman should go switching off a random X chromosome in her cells, but in fact this "X inactivation" is probably essential for her survival. As early as the 1930s biologists had identified a potential problem for women that is a direct result of the X-inequality between the sexes. Men are living proof that human bodies cope perfectly well with exactly one X, and many geneticists had begun to worry about how women manage to survive with two active X chromosomes. Why are women not overdosed with X gene products?

To the casual observer, this may look like a case of scientists worrying about something that is of no real concern, but there was a reason for their chromosomal consternation. That reason was that many of the most common genetic diseases of the human race occur because babies inherit extra chromosomes, or even just extra fragments of chromosomes. For example, children with an extra chromosome 21 have Down syndrome, and inheritance of extra copies of most other chromosomes is usually fatal long before birth. Acquiring an excess chromosome is a serious, and often deadly matter—and this is why women's X chromosomes presented such a paradox. The X is the only chromosome that is present in a single copy in some healthy people (XY men) and two copies in others (XX women), so why do women show no apparent ill effects?

But once again Mary Lyon's theory comes to the rescue—X inactivation provides the answer to women's X chromosome problem. To avoid being overdosed with X genes, the female embryo simply inactivates an X chromosome in every cell in her body, so that each of her cells contains two X's, but only one functional one. Women are left with the same number of functional X chromosomes as men—one. So, in short, X inactivation has a very important function: to protect women from the toxic effects of their own chromosomes. Women have, in effect, become inherently mixed creatures so that men can cope with just one X chromosome. Indeed, you can think of female mosaicism as the other side of the coin from sex-linked diseases in men—it is women's attempt at coping with having two X chromosomes, whereas sex-linked disease is evidence of men's failure to cope with having just one.

There are, of course, more ways than one to skin a calico cat. Animals have been confronted by the potential problem of X overdose many times during evolution—and in fact, any species that has XX females and XY or X– males faces exactly the same dilemma. Just because mammals avoided it by X inactivation does not mean that there are no other options. Roundworms, for example, do something slightly different. Instead of female worms switching off one X and keeping the other one functioning, they simply reduce the activity of both chromosomes by half. Flies achieve the same effect by doing the complete opposite. Female flies are allowed to use both their X chromosomes, but males simply double the activity of their lone

X. Controlling rogue surplus X chromosomes is obviously a common problem, but it does not seem to matter how animals actually achieve it.

One of the most challenging aspects of X inactivation is the question of how women manage to do it. Quite simply, no other chromosome ever just switches itself off in such a wholesale manner. The individual genes that they carry may flicker on and off throughout life, but all other chromosomes remain, on the whole, open for business. Because the X is unique in this respect, there is nothing with which to compare it—no similar chromosome to help us.

Right from the start, geneticists realized that the process of X inactivation is not just a simple switch. Instead, every cell in a female embryo must go through a series of carefully choreographed steps to achieve it. First of all, female cells must somehow "count" their X chromosomes to establish that there are in fact two—and we know this because XO women with only one X do not undergo X inactivation. Next, every cell must somehow "choose" which X is to be inactivated and which is to be spared—a process that occurs apparently at random. And finally, the switching off of one X and the switching on of the other must occur, and it must take place in such a way that it is perpetuated throughout the descendants of each cell, to yield a patch of cells in the adult woman that share the same active X.

Working out the intricacies of these counting, choosing, and switching processes seems like rather a tall order, and indeed a considerable amount of labor has been required to tackle the problem. Yet within the last decade, geneticists believe they have worked out the gist of how women actually achieve X inactivation. And remarkably, the whole process can probably be traced back to a single gene. A clue to the location of that gene is given by the pattern in which an X chromosome, when marked out for a life on the sidelines, gradually crumples down into a Barr body. That crumpling starts at a particular point on the X (see arrow in figure below). That part of the chromosome crumples, and then the crumpling seems to spread along the X in both directions, rather like falling lines of dominoes, until it reaches the chromosome's tips. Clearly this "X inactivation center" must have a very important role in the whole process.

Geneticists' interest was piqued further when a gene was found behaving rather strangely on this part of the inactivated X chromosome. It was, in short, not inactivated, but churning out large amounts of its product. And conversely, the copy of this gene on the "active" X chromosome is inactive, as is the copy on a man's lone X. In other words, this gene only seems to function on an inactive X chromosome, giving it the name *Xist* (inactivated X chromosome–specific transcript). The suspicious behavior of *Xist* led its discoverers to claim that this is the very gene that switches an X chromosome off—hence its rather contrary behavior.

Other studies have confirmed that *Xist* is the key to splitting women into two unequal halves. No matter how many supernumerary X chromosomes are stuck into cells, *Xist* reliably switches on in all except one. Its powers are not even restricted to the X chromosome, so that if *Xist* is spliced onto another chromosome, then it can inactivate most of that, too. Finally, and most intriguingly, genetic engineering wizards are now able to disrupt different parts of *Xist* and have even shown that some regions of the gene are involved in the X counting process, others in choosing which X is to be doomed to inactivation, and others in the inactivation process itself. So this single gene lies at the heart of the innate complexity of women. This is the stretch of DNA rung-code that forces one half of the female body to use one X, and one half to use the other. And all this to stop women being overdosed by the very chromosomes that made them female in the first place.

Exceptions that Prove the Rule

Some of you may have been left wondering whether this blasé deactivation of X chromosomes affects how they are passed on to the next generation. After all, throughout this book I have been telling you that women can bequeath either of their X chromosomes to each of their children. So what happens to the children who get the inactivated one? Well, there is a neat little trick going on inside every women to make sure that this never happens.

When I said that every cell in a woman's body uses just one of her X chromosomes, I was not quite telling the whole story. In fact, there are several thousand cells in the female body that switch the Barr body back on—and those cells are the eggs sitting expectantly in the ovary. The primordial germ cells that give rise to eggs do not start their life in the ovary. Early on in a girl's embryonic development, they migrate from a sac hanging off the developing gut and invade the embryonic ovary. As they do this, a remarkable thing happens—their *Xist* switches off and their Barr bodies dissolve once more into fully active X chromosomes. Their X is somehow "un-inactivated" and made pristine, ready for transmission to the next generation. A mother must not pass on an already inactivated X chromosome, since each egg has no way of knowing whether it will form a girl or a boy, and a boy would perish without an active X. So in the human life cycle, X inactivation follows an interrupted pattern—every X chromosome is inherited without any hint of previous inactivation, so that every daughter must start the inactivation process afresh for herself.

So among all the myriad cells of the female body, eggs are special. Yet this specialness is not without its problems. Eggs have two fully functional X chromosomes, which of course is exactly the predicament that X inactivation was supposed to avoid. If women go to so much trouble to protect cells by switching off an X, then how can their eggs survive with both of them switched on? It is not as if this is just a transient problem, either—once eggs are formed in an embryonic girl, they may have to survive for over four decades before they get a chance to be fertilized and make a baby. And all this time they have two active X chromosomes sitting uneasily side by side.

Geneticists simply do not know how eggs cope with the potential problem of X overload. Perhaps there is just something about being an egg that means that X overdose does not matter—just as it does not seem to matter in very early embryos before X inactivation has ever taken place. Alternatively, eggs may have some other, enigmatic method of avoiding the adverse effects of excess X chromosomes, but that does seem to be making matters inordinately complex—why can they not simply use the same system as all other cells? Ironically, eggs are one of the few cells that only really need half of their chromosomes. Just after fertilization, they dispose of one-half of their genetic material, including a whole X chromosome. The reason they do this is, of course, that they only contribute a half-complement of genes to the new life they are helping to create because a sperm contributes the other half. Why eggs have to wait a few decades until they are eventually fertilized to dump their unneeded X, instead of doing it as soon as they form, is a mystery. After all, early disposal of an X would mean that they would not have to survive so many years under the constant threat of X overload.

As it turns out, eggs are not the only strange exception to our ideas of X inactivation, and there is now evidence that the random pattern of X inactivation proposed by Mary Lyon is not quite as universal as we once thought. Geneticists had just become reconciled to the idea that women are an equal mixture of two types of cell—one using their mother's X and the other using their father's—when a series of discoveries rather upset the X inactivation apple cart. The first challenge to the idea that women are a mixture of two equal parts came from a strange quarter: marsupials. As I mentioned in Chapter 1, our pouched cousins decide their sex in very much the same way as us, with XY males and XX females. This means that female kangaroos face the same problems of X overdose as women, and indeed they also solve it by inactivating a single X chromosome in each of their cells early in embryonic life. However, the way in which they achieve this is very different. Instead of each embryonic cell picking an X at random, all the cells simply switch off the X chromosome inherited from daddy kangaroo. So female marsupials never use their father's X chromosome at all—it languishes unused in most cells in their body, as a Barr body. Just as in women, however, the inactivated X is reactivated in marsupials' eggs, just in case it needs to be passed on to the next generation.

The upshot of all this is that although female marsupials switch off an X, they do not end up as a mixture of two different cell populations. So unlike women, they are not mosaics. Although Mary Lyon's theory is correct for people, it is not the only way for females to

X-inactivate their way out of trouble. In other words, the randomness of X inactivation is not an essential part of the story—humans have it, but some other mammals do not.

Of course, this raises the question of whether the process of switching off women's X chromosomes is really random at all. What do we find when we conduct a census of all the cells in a female body—do they really use the X chromosomes in a neat fifty-fifty ratio? For obvious reasons, carrying out this census is not really practicable in humans, so instead it was done in mice. And indeed, X inactivation does not quite divide the murine female's body into nice equal paternal-X and maternal-X domains. Instead there is a slight but very real bias. Because slightly more cells switch off the X that the mouse inherited from her mother, female mice make slightly more use of their father's X than their mother's.

Although this slight preference for maternal X inactivation could be ascribed to some trivial failure of chromosomal even-handedness, there is a particular type of cell in which the bias is far more marked, and intriguingly the bias is in the other direction. These cells always inactivate the paternal X and use only the maternal one. I find this dramatic exception to randomness in X inactivation the most exciting of all, and that is because my own line of research is into the interactions between pregnant mothers and the babies growing within their wombs.

All growing mammalian fetuses are surrounded by a complex arrangement of membranes that not only protect them, but also form the placenta through which they are nourished. Although these "fetal membranes" are essential for the survival and growth of unborn mammals, they are also a potential Achilles' heel. Because they are the only part of the fetus that is directly exposed to the mother's own tissues, these membranes are the interface at which the developing infant may be recognized as "foreign" by its mother's immune system and rejected, rather like an organ transplant. Despite the fact that a baby inherits about half of its genes from its father, and many of those may yield products foreign to its mother, unborn babies still do not seem to get "rejected" by their mothers.

This is why people like me find it so fascinating that it is the cells in these fetal membranes which preferentially switch off the paternally inherited X. What, exactly, are these cells trying to hide from the mother? It is, of course, tempting to suggest that baby girls are plotting some sort of clever subterfuge to prevent themselves being recognized as an invading "alien," but there is not yet much evidence to support this. Because of our attempts at organ transplantation, we know quite a lot about how bodies reject foreign tissue. For example, we know that there is a special family of genes which induces most graft rejection, and we know that none of these genes are on the X chromosome. So, from that point of view, switching off the paternal X in the fetal membranes would not really seem to achieve much for a growing baby girl. However, the jury is still out on this one, and we cannot yet rule out that unborn girls use biased X inactivation in their membranes to protect themselves from maternal attack. (This is not, however, an issue for boys, as they only have their mother's X anyway.)

I hope I have convinced you that avoiding X chromosome overdose lies at the heart of what it means to be female. Because men have to cope with just one X, it seems that women must be designed to function in the same way. To allow this, one X is switched off early on in a baby female's development—either at random as in girl's bodies, or by suppressing the paternal X in marsupials' bodies and unborn girls' fetal membranes—and then re-activated in the egg cells in the ovary.

However, all this jiggery-pokery with the way that animals use the X also seems to have left that chromosome in a strange, almost isolated state. This predicament has resulted from the fact that

mammalian cells now expect there to be only one active copy of genes
borne on the X chromosome, whereas there are two active copies of
every gene carried on all the non-sex chromosomes. Because of this, it
seems that every gene on the X has to "work twice as hard" in some
way. There is not usually an active duplicate copy of an X-borne gene
present in the same cell—males' cells contain no other X, and females'
cells will have inactivated it.

So X genes must be set to be twice as active as genes on a non-sex
chromosome. Strangely enough, this has created a kind of barrier
between the X chromosome and all the others. Over the course of
evolution, genes flitted freely around the non-sex chromosomes, but
they do not seem to like moving from a non-sex chromosome to the
X. Presumably, this is because an animal which inherits a gene that
has made such a move will find that this gene is only half as active as it
needs to be, so the animal will be disadvantaged in the fight for survival. And the converse will also be true—genes chopped off the X
and glued to a non-sex chromosome will suddenly seem over-active.

In other words, the process of X inactivation has isolated the X
chromosome. Genes on non-sex chromosomes seem loath to immi-
grate into the X, and X chromosome genes find it similarly difficult to
emigrate. Because of the idiosyncratic way the X has managed its
affairs, it has become ostracized by the other chromosomes. In many
ways, this could be seen as the last phase of that sad divorce of X and Y
that I mentioned in Chapter 1. You may remember that the X and Y
are thought to have evolved from an amicable pair of non-sex chromosomes that became estranged after one of them acquired the gene
which determines babies' sex. They stopped swapping genes with each
other, and the chromosome with the sex gene (the Y) gradually shed
most of its other genes until it became a shadow of its former self.

Now it would seem that this whole evolutionary process has led to
its estranged partner, the X chromosome, having to resort to complex

tricks so that one sex can survive with one X while the other survives with two. In turn, these tricks have meant that the X has ended up not only divorced from its old partner, but also isolated from all the other chromosomes as well. And this isolation is probably the reason why the X has survived while the Y has withered—the X is simply incapable of losing its genes to other chromosomes, and so they languish on this stubborn chromosomal island, unable to escape. Even though they render males vulnerable to all those awful sex-linked diseases I discussed in Chapter 2, still the X cannot offload its vital genes to somewhere less precarious. The divorcée has been cast out of its own community, and is finding life hard on its own.

Some You Win, Some You Lose

Identical twin girls are never as identical as identical twin boys. Neither as identical, nor as rare.

In 1977 there appeared a strange report of a little girl with Duchenne muscular dystrophy. In the last chapter, I explained how this disease is far commoner in boys than girls, and that girls are only really likely to get it if they are the product of some form of incest. Yet there was no hint of an incestuous cause of this girl's disease, but instead there was another clue which suggested that something far more remarkable than simple genetic accident had caused this child's disease. She had an "identical" twin sister, and that twin was unaffected by the disease.

How could this be? Duchenne is very clearly a genetic disease, and identical twins are supposed to inherit the same genes. So at first sight it is hard to see how the girls could be born with such different fates. At first, geneticists suspected that some chance error in the chromosomal dance had occurred—they hunted for evidence that the girls had somehow inherited different, abnormal complements of

X chromosomes, or even that one of their chromosomes had acquired or lost a piece. And yet all the evidence pointed to them being absolutely *bona fide* "identical" twins. They shared blood groups, tissue types—everything.

Since those original twins, similar pairs of so-called discordant twins have been reported—one with an X-borne genetic disease, and one not—and as methods of gene analysis have become more and more sensitive, again and again they have been found to be true "identical" twins. Because they have formed from a single fertilized egg that subsequently splits, their cells contain almost indistinguishable chromosomes and genes. Today, these girls can even be shown to be carrying exactly the same form of damage to the dystrophin gene on one of their X chromosomes. They are genetically similar enough for every last DNA ladder rung of their huge dystrophin genes to be the same. Yet these so-called "identical twins'" fates diverge.

The differences between these "discordant" identical twin girls can be extremely dramatic—muscular dystrophy discordance can lead to the birth of twins, one of whom is destined to become a world-class athlete, while the other will be wheelchair-bound. We may have become resigned to the heartless way in which nature deals some people a bad genetic hand, but the idea that such unfairness can be inflicted on identical twins seems almost perverse in its paradoxical callousness. Yet there is one additional unfairness that gives us a clue to why "identical" twin girls can differ in such a spectacular way—identical twin boys simply never do. And what do girls have that boys do not? Two X chromosomes.

Somehow, "identical" twin girls can differ considerably in the extent to which they use their two X chromosomes—even though those chromosomes are the same in each twin. And the randomness of Mary Lyon's X inactivation is the key to these differences. Because an embryonic baby girl switches off an X at random in each of her cells, there is no way of predicting which bits of her adult body will use which X. So the pattern in which the paternal X and the maternal X are used will not be the same in two "identical" twin girls, and this is how they can become discordant for genetic diseases carried on an X chromosome.

Girls can easily inherit a damaged dystrophin gene (usually from their healthy mother) and a normal dystrophin gene (usually from their healthy father). Yet if an unusually large number of their future muscle cells use the X carrying the damaged gene, the girl may show signs of the disease, even though she has a perfectly normal X. This normal X will be no help at all if it is inactivated in the cells that count—the muscle cells. The double nature of women is coming back to haunt them. Their eventual health is not determined simply by the genes they inherit and their environment—it is also dictated by the random process of X inactivation that takes place a few days after they are created.

Approximately one-twentieth of a girl's genes reside on an X, so you can see that identical twin girls can potentially differ by up to 5 percent in the genes they use just because they use their X chromosomes differently. They frequently do differ quite markedly in their X usage, often to the extent where the term "identical" twins does not really make much sense any more. Indeed, scientists now use the term "monozygotic," which simply means that the twins are derived from a single fertilized egg. So X inactivation explains why identical twin girls are so much less similar than identical twin boys. There are other reasons why so-called "identical" twins of either sex are never truly identical, such as minor genetic changes that take place after they split apart, but the differences caused by different patterns of X inactivation are the most dramatic.

While discordance of X-linked genetic diseases can consign twin girls to such arbitrarily different fates, it may also provide them with a unique opportunity for treatment. Quite simply, the twin with the disease has a potential source of salvation not available to males with the disease—a healthy, genetically almost identical twin. In the case of muscular dystrophy, it is possible that the healthy twin could act as a donor of healthy tissue which uses the undamaged X. What is more, the fact that identical twins share the same tissue type means that transplantation can be carried out without any of the usual risks of graft rejection. Exactly this approach has already been attempted by a Canadian team. They found a pair of monozygotic twin girls, one with symptoms of Duchenne muscular dystrophy and one without, and took a biopsy sliver of muscle from the healthy sister. From this sliver, they succeeded in growing little colonies of immature muscle precursor cells in the laboratory. By some clever gene analysis, they could identify which of these colonies were using the undamaged X chromosome, and injected them into the arm muscles of the sister with muscular dystrophy. The results were not miraculous, but they were nonetheless exciting. One of the injected muscles showed a small but very real increase in its ability to straighten the wrist. Unsurprisingly, there was no evidence of any rejection of the transplanted cells, and indeed biopsies taken one year later showed that the sisterly cells had happily set up home in their new location. Although it is a long way from a whole-body cure for muscular dystrophy, this rather unusual form of genetic therapy shows that new and exceptional ways of acquiring diseases can often lead to new and exceptional ways of treating them.

Yet the existence of these discordant twin girls can really be seen as a rare natural reminder of just how deeply ingrained the mixed nature of women actually is. The fact that all "normal" women are made up of a random mixture of two sorts of different cells, each using a different X chromosome, is not just an interesting piece of biological trivia—it is something integral to our modern concept of femaleness.

Nowhere is this more apparent that in sex-linked diseases. You may remember that when I was discussing hemophilia and muscular dystrophy in the last chapter, almost as an aside, I mentioned that women in families afflicted by sex-linked diseases are often slightly affected themselves. Even though they rarely show overt symptoms, the blood of women with a single damaged factor VIII gene may clot more slowly than that of normal women. Similarly, women with a damaged dystrophin gene on one X chromosome may have higher levels of creatine kinase—the classic indicator of muscle damage.

Now that geneticists understand X inactivation, such symptoms seem quite logical. For example, factor VIII is manufactured in the liver, and men with a defective factor VIII gene lack an important component of their blood clotting system, leading to hemophilia. Women with a functional copy of the gene on one X chromosome and a defective copy on the other X are in a different predicament, however. Because of X inactivation, half of their liver uses one X and the other half uses the other X. So half of their liver will be able to make factor VIII and the other half will not. In everyday life this does not seem to make much difference—these women do not experience the symptoms of hemophilia, presumably because there is a safety margin built into the clotting system and a half-dose of factor VIII is quite sufficient for normal life. Or maybe the "normal" half of the liver can slightly over-produce factor VIII to compensate for the inability of the other half to produce it. Whatever the mechanism, sex-linked diseases that affect body processes with a degree of flexibility are not much of a problem to women with just one damaged X.

Things are slightly different with muscular dystrophy. Whereas liver cells simply pour factor VIII into the circulation to be used elsewhere,

muscle cells make dystrophin for their own use. And this means that
the approximately fifty-fifty split of normal and dystrophic cells can
actually be evident within carrier women's muscles themselves. Women
with elevated creatine kinase levels sometimes have a very distinctive patchy pattern to their muscles. Groups of normal cells are interspersed with groups of dystrophic cells in a mosaic arrangement.
In discordant twins, of course, one or other type of muscle cell
predominates.

This calico pattern in women becomes more evident in other sex-
linked diseases in which actual location of diseased cells becomes
more important. One liver cell may be able to make up for another's
failing, and even in muscle the strong may be able to compensate for
the weak, but this is not always the case in other tissues. Probably the
most dramatic example of this is a disease called "anhidrotic ectodermal dysplasia," which is doctor-speak for "developmental abnormality of the outer layer of skin in which there is not enough sweat." This
is an X-borne, sex-linked disease that results in failure of the skin to
develop sweat glands. Now skin is very visible, and one bit of skin
cannot really take over the role of another bit of skin, with the result
that women with one gene for anhidrotic ectodermal dysplasia are
visibly patchy. Half of their skin has sweat glands and the other half
does not—they have become a visible antiperspirant map of the random pattern of X inactivation that makes up every woman. They are
the calico cats of the human world.

Although often hidden, the disease for which women most often
show a patchy pattern is color blindness. Just as there are many men
out there who are color blind, there are also many women who are
"half" color blind. Most women who carry a single gene for red-green
color blindness—and there are lots of them—have a patchy, mosaic
pattern of photoreceptor cells in their retina. Approximately half of
the retina contains "normal" distinct red and green cones, but the

other half does not. Just like any other part of the female body, the retina is made of a mixture of two types of cell, each using a different X chromosome, and these cells are arranged in random patches. However, the fact that half the retina is color blind does not seem to worry the brain very much—the supercomputer in women's heads has little trouble using the color information from the red-green functioning half of the retina to fill the informational deficit from the other half. In fact, fairly advanced methods are required even to detect this patchy color blindness, such as directing red or green lasers onto tiny little clumps of photoreceptors in women's eyes.

So women's bodies truly are mixed—in a very real way that springs into relief whenever an X chromosome is damaged. Each woman is one creature and yet two intermingled, as it were. Yet this spectacular but little-mentioned wonder of nature may have still more to tell us about the origins of the female of the species.

You may have noticed a slight inconsistency in what I have written about X inactivation so far. I was keen to emphasize how a woman's body is divided into two roughly equal portions, each using a different X chromosome. But I also claimed that identical twins can be discordant for a particular sex-linked disease because each preferentially inactivates a different X chromosome. How can most women be assumed to be approximately fifty-fifty mixtures, while we simultaneously claim that some female identical twins show a dramatic bias toward use of a particular X? This is, in fact, a topic of much current debate, largely because it may explain the mysterious mechanisms underlying twinning in humans.

Twinning has long gripped the imagination of the inquisitive. To the prescientific mind, the fact that most human mothers bear their children one at a time, yet a few produce them in batches of two or even more, must have been one of the great mysteries of nature. In many societies, twins have been treated as exceptional—and not always to their benefit. In medieval England, for example, twins were considered not just unusual but frankly diabolical. Also, fraternal twins were often thought to have been sired by different fathers, which often led to one being killed to preserve the reputation of the mother. Clearly no one wanted to compare human reproduction with that of farm animals or pets, or they would have seen how wrong they were.

We now know that there are two different forms of twinning. Fraternal (nonidentical, or dizygotic) twins occur when a woman produces two eggs and both are fertilized, leading to the birth of two children as dissimilar to each other as any other siblings would be. Fraternal twinning can "run in families," because the propensity of women to ovulate twice is a partially genetic characteristic that can be inherited. In contrast, monozygotic (so-called "identical") twins result when a single fertilized egg splits into two as it develops. Monozygotic twinning is rarer, not inherited, and poorly understood, at least in humans. Some mammalian species have taken to it with gusto, however—nine-banded armadillos usually produce litters of identical quadruplets.

For some time, reproductive biologists have pondered a strange but not widely reported fact about "identical" twinning—it is a predominantly female phenomenon. More female monozygotic twin pairs are born than male, and intriguingly this trend seems to be linked to the point at which the developing embryo splits into two. If the embryo splits quite early, then the two babies are entirely independent, each having its own full set of membranes. Roughly 51 percent of these twins are female—not a staggering bias, I will admit, but more is to come. If the split occurs slightly later, then the twins may end up sharing their outer membrane, and 57 percent of these are female. If they split slightly later again, they must share their inner membrane too, and 70 percent of these are girls. Finally, if the embryo splitting occurs so late that it is incomplete, then conjoined twins result, and 75 percent of these are female. Just to be contrary, conjoined twins are also called "Siamese," after a particularly famous set of Thai conjoined twins who were, bucking the trend, male.

So why is there a creeping dominance of "identical" twinning of girls? Obviously male embryos do split, but they do it less often, and earlier. So little is known about the process of monozygotic twinning that developmental biologists have attached great significance to the two things they do know—female embryos split more often and later, and the resulting twins often preferentially inactivate different X chromosomes. In fact, some researchers claim that all female monozygotic twins show oppositely biased use of their two X chromosomes, but this is contentious.

A challenging theory has been developed in an attempt to unify and explain these two phenomena. It has been suggested that once a female embryo has randomly inactivated her X chromosomes, and thus subdivided herself into two different populations, these populations may then actively repel each other. The idea is that most of the cells that use the father's X clamber away from the cells that use the mother's X, and vice versa, and that this can eventually lead to the embryo splitting into two. The two resulting embryos will show a strong bias toward one or other X, and this is why some female "identical" twins end up discordant for sex-linked diseases. This theory has even been extended further to claim that any singleton girl born with a strongly biased pattern of X chromosome inactivation probably had a monozygotic twin that died in the womb.

This theory is a challenging one, as it is the first to give us any insight into the mysterious world of monozygotic twinning. Of course it cannot explain all such twinning—it cannot account for male "identical" twins, for a start—but the idea that cells in an embryo can repel each other enough for it to be broken asunder is really rather charming. And surely the fact that female identical twins often split around the time that X inactivation is occurring cannot be a complete coincidence? How strange, then, that monozygotic twins, perhaps wrought in an act of mutual repulsion, often feel such a special bond to each other in later life.

If the unequal use of X chromosomes in discordant female twins seems dramatic, I should add as a footnote that there is another situation in which a woman's cells can act in a very unequal way with regard to their X chromosomes. Not content with encroaching on every other part of our life, the X has also explained to us how cancer develops.

By the time most tumors are discovered, they are usually quite large, and consist of millions, if not billions of cells. In the early days of cancer biology, no one really had any idea how cancers actually start—whether they were the copious progeny of a single deranged progenitor cell, or whether they formed from a patch of tissue made up of many cells, each activated by the same local cue. This question has proved to be crucial in the fight against cancer. Only by knowing how tumors form can we hope to stop them gaining control over us. It was the X chromosome that answered this conundrum for us, when it was discovered that however large a woman's tumor has grown, and however far it has disseminated around the body, it always has the same inactivated X chromosome.

Although women are mosaics, their cancers never are, and this is because they are never derived from more than one cell. If they were, then some women's tumors would be mixed, containing cells with different inactivated X chromosomes. All tumors start as an single act of sinister madness in a lone, fatal cell. If we can stop that single cell, then we can stop cancer.

The Historical Origins of the Civil War

Although studying women's tumors taught us the origins of cancer, women are not inherently any more susceptible to cancer than men. Researchers studied them simply because they happen to be the sex that undergoes X inactivation. Yet there is a different type of disease—sometimes mild and sometimes devastating—to which women are more susceptible than men, and X chromosome inactivation may explain this susceptibility. These are the autoimmune diseases.

As its name suggests, autoimmune disease occurs when the body's own immune system—the array of cells meant to fend off microbial invaders—starts to attack some innocent part of the body itself. The effects of this onslaught depend on which body part is attacked—muscle, nerve, insulin-producing cells, thyroid, skin—but the basic process of immune self-mutilation seems to be the same. It may seem completely stupid for immune cells to start attacking part of the body to which they belong, but as I hope to explain, it can be an easy mistake to make. In fact, in some ways it is surprising that it does not happen more often.

One of the central players in the immune system is a little cell, usually found in the blood, spleen, or lymph nodes. This is the lymphocyte: small, round, and featureless, but just about the cleverest little cell in the whole body. Lymphocytes have a tremendous responsibility thrust upon them. The body is continually threatened by marauding bacteria, viruses, parasites, and fungi, and although they can often be fended off by relatively simple means, many do manage to break through the defenses and gain entry into the body. Once established, it is the job of lymphocytes to recognize these invaders as foreign and coordinate a ferocious counterattack.

A lymphocyte's job may sound relatively straightforward, but there is a catch. Life is just too unpredictable. Lymphocytes must be able to recognize invaders that the body has never encountered before, and even invaders that no other member of the animal's species has ever had the misfortune to meet. It seems almost inconceivable that any sort of cell could be clever enough to recognize an entirely novel microbe as foreign, but this is exactly what lymphocytes do, throughout your life. Obviously they cannot be preprogrammed with all the possible invaders that they might meet, because a microbe that has never been encountered before is, obviously, an unknown quantity.

Every day, millions of new lymphocytes are churned out by your bone marrow, and each one is able to bind to just one sort of chemical. They then decamp to the thymus, a pale soft organ in the chest, just under the breastbone. In many ways, the thymus is a school for lymphocytes, where they undergo what is known as "thymic education." This means that specialized teacher cells present them with all the chemicals that they are ever likely to encounter in the entire body. That may seem quite a feat, but it is nothing compared to the level of discipline exacted on the young lymphocyte students. If any of them recognize a chemical that is a component of the animal's own body, then they are summarily destroyed. So, in common with many harsh educational establishments, the thymus has a very high dropout rate, but the students who do stay the course are extremely highly trained. By the time they exit the thymus, mature lymphocytes should be able to recognize just about everything except components of the animal's own body.

So the lymphocyte is the cell that makes the final decision about what is part of the body and what is foreign. It usually makes this decision correctly. Well trained in the thymus, lymphocytes very rarely attack normal parts of the body. But on the rare occasions when they do, the concerted onslaught meant for invading microbes is launched instead on innocent body cells, usually leading to their destruction. This is autoimmune disease—rheumatoid arthritis, multiple sclerosis, lupus, and some diabetes. All these diseases result from a lymphocyte's rare act of misidentification. Once a lymphocyte makes this disastrous mistake, there is little that can be done to prevent the carnage. Spurred on by its apparent discovery of alien material, the lymphocyte divides repeatedly until it has formed a veritable army of misguided progeny, all hell-bent on destroying the hapless target. As you can probably tell from that very truncated list of diseases above, autoimmune disease is immensely widespread, affecting perhaps one-fifth of people during their lifetime. Not surprisingly, a tremendous research effort is underway to find out just how the immune system goes so badly off the rails.

The reason I have brought up the subject of autoimmune disease is that it has a distinctly misogynistic streak. Multiple sclerosis, for example, affects twice as many women as men. Lupus affects women ten times more often. And remarkably, Hashimoto's thyroiditis occurs fifty times more commonly in women, even though women's and men's thyroids are pretty much indistinguishable. All in all, maybe four-fifths of those with autoimmune disease are women. Why should this be? Why are women's lymphocytes so much more likely to go on these erroneous internal rampages?

There are two snippets of information that give us a clue to why women get autoimmune disease more often. The first snippet is that some autoimmune diseases can follow a rather strange pattern in some women. Some of them, especially the ones that affect the skin, can be seen to trace out a patchy pattern—not dissimilar to the mosaic pattern of anhidrotic ectodermal dysplasia or calico cats. The second comes from a small group of men who are born with an abnormal complement of sex chromosomes. You may remember from Chapter 1 that men are occasionally born with two X chromosomes and a single Y—indeed these XXY men are part of our evidence that it is the presence of a Y rather than the number of X's that decides the sex of a human baby. They are relevant to our present story because, intriguingly, XXY men are more prone to autoimmune disease than XY men.

You can probably tell which way this line of argument is going. What do XXY men and XX women have in common, apart from being more susceptible to autoimmune disease? Two X chromosomes, of course. And why do occasional autoimmune diseases trace out a patchy mosaic-like pattern? Could they by any chance be attacking just half of the female body? This enticing information has led many immunologists to suspect that women are more prone to these diseases as a strange side-effect of the process of X inactivation. They have proposed that the division of the female body into two factions can sow the seed of a strange sort of civil war, in which the immune system somehow mistakenly attacks cells that use a particular X chromosome.

At this point, the X starts to look rather like a feminine liability. Because it was unable to transfer its genes to safer, non-sex chromosomes, the two different populations of cells that make up a woman's body may well be making subtly different proteins from the genes on their different X chromosomes. Because of this, these two groups of cells probably even "look" slightly different to the lymphocytes that constantly patrol the body, searching for invaders.

Yet the double nature of women alone is not enough to explain autoimmune disease. The whole point of the process of the education of lymphocytes in the thymus is that they get exposed to every protein made in the body, so why should this careful training not also work in women? If the thymic teacher cells are drawn from both of a woman's body's populations of cells, then you would expect all lymphocytes to be carefully trained not to attack cells regardless of what X chromosome they use. And this seems to be exactly what happens in the majority of women who never get an autoimmune disease—lymphocytes are taught from an early age that the proteins made from the genes on both X chromosomes are not legitimate targets.

But what happens if the thymus does not show its lymphocyte students the protein products of both X chromosomes? If a developing girl has a strongly biased pattern of X inactivation, then it is perfectly possible that most, or even all, of the cells in her thymus will use the same X chromosome. And if that happens her lymphocytes will not be exposed to the products of the other X. When these lymphocytes enter the circulation they may well encounter small clumps of cells that use the other X, and suddenly recognize some of their proteins as foreign. Then the civil war starts.

So it really does seem possible that the two halves of a woman's body are so different that the immune system can mistakenly think that one half is a foreign organism. This, of course, would explain why XXY men have the same increased incidence of autoimmune disease, and why these diseases occasionally seem to trace out a mosaic-like pattern in women. Perhaps I have uncovered the flip-side to all those problems that men have with their vulnerable lone X chromosome—women's continual internal battle to ensure self-recognition.

No one really knows how many women have autoimmune disease because they use their X chromosomes unequally in the thymus, but I would suggest that even this innate X bias may not be necessary to explain female susceptibility to these diseases. As lymphocytes pass through their thymic schooling, they do not really come into contact with many teacher cells. Such things are, of course, very difficult to count, but I do not think that twenty teacher cells is too miserly an estimate. Even if the thymus is a neat homogenous fifty-fifty mix of cells using the paternal X and cells using the maternal X, what are the chances that a lymphocyte will meet twenty random teacher cells, all of which have inactivated the same chromosome? By my calculations, the chance of this happening for each lymphocyte are one in 524,288.

These may seem like long odds, but when you consider that billions of lymphocytes enroll in the thymic school in a woman's lifetime, it suddenly seems very likely that many will never be exposed to the proteins made from the genes on one of the two X chromosomes.

So perhaps the preponderance of autoimmune disease in women does not have to be explained by bizarre skews in the pattern of X chromosome usage. Maybe an insufficient variety of teachers is enough to explain women's predicament. One thing I have wondered for some time, and have never been able to find out, is whether female marsupials are more prone to autoimmune disease than males. As a veterinarian, I am continually made aware that domestic animals are just as susceptible to autoimmune disease as humans, so why not marsupials? You may recall that female marsupials are not mixed "mosaics"—they simply inactivate the paternally inherited X chromosome in almost every cell in their body. So, if the theory is correct, then they should be no more susceptible to these diseases than male marsupials.

I must stress, however, that there are competing theories that also attempt to explain why women get autoimmune disease more often than men. Maybe female reproductive hormones, or the hormones produced by the baby during pregnancy, somehow drive the immune system to error. Even more dramatically, it is known that cells usually seep out of babies into their mothers' bloodstream, and some of these are thought to survive for decades in the maternal bone marrow—could these alien interlopers be the cause of the disease? This area is a subject of great debate at the moment, but no theory is perfect. For example, all the theories that claim pregnancy or female hormones as the cause cannot explain why the disparity between males and females is already apparent by puberty.

Scientists still have a long way to go before they discover exactly why women are more likely to get autoimmune disease, but hopefully when they do, it will help them explain why anyone gets these strange diseases at all, and why evolution has allowed one sex to become more vulnerable to them. One thing I can assure you is that what is already known about autoimmune disease is far more complex than I have suggested. If you want to tax your brain further, you may want to consider what happens on the rare occasions when the immune system starts attacking the Barr body itself.

There Is Always Another Way

Before I finally let you escape from the clutches of X inactivation, I should tell you that all this mosaicism and chromosome switching has one more major effect on us—on our sexuality itself.

I hope that I have not given you the impression that X inactivation is a flawless system that allows men to be men and women to be women, and never the twain shall meet. What with all the trouble it causes, you might expect that X inactivation could at least get its main job right—allowing the human population to be divided into XX women and XY men, both of whom can survive with the X chromosome bounty that nature has given them. But nothing in biology ever seems to work perfectly, and this is especially true for X chromosomes.

In Chapter 1 I proposed that the Y chromosome does not really amount to much at all—a shriveled little thing. Because of this, babies can be born with an abnormal number of Y chromosomes—two or more—and suffer few ill effects. Certainly, the Y chromosome is unusual in this respect, as an excess number of any non-sex chromosomes is usually lethal long before birth (Down's syndrome is an exception to this). Yet rather like the Y, an abnormal number of X chromosomes is also often compatible with life. Compared to the strict requirement for the correct number of most chromosomes, the human body is rather lax about the number of sex chromosomes it needs, and in the case of the X, the reason for this relaxed approach is X inactivation.

In 1938, Henry Turner described an unusual set of symptoms that he had discovered in some adolescent girls. The girls appeared rather short and sexually immature. They had some distinctive characteristics—some had flaps of webbed skin between their shoulders and neck, others had an arched palate, and still others had low-set ears, occasionally accompanied by deafness. Some had heart defects and others went on to develop diabetes and thyroid problems. Although these features did not really seem to have very much in common, they cropped up with moderate frequency as a well-defined group— Turner's syndrome.

When cells from these girls were squashed, stained, and examined under the microscope, the cause of the syndrome eventually became apparent. These girls usually have forty-five chromosomes, instead of the usual forty-six, and the chromosome they lack is an X. Girls with Turner's syndrome usually have twenty-two pairs of non-sex chromosomes and a single X—there is no other X and no Y. These are, in fact, the "XO" girls I have mentioned occasionally throughout this book, and they make up perhaps 0.02 percent of all human females. They certainly look female, which we would expect, as we have seen that the presence of a Y determines whether an embryo is a boy or a girl.

The more surprising thing about Turner's syndrome is that these girls ever get born at all. After all, any embryo that is missing a non-sex chromosome is doomed to a very short existence, so how do embryos with Turner's syndrome bluff their way through to birth and beyond? The answer is, of course, that human cells are quite happy to function with just one X. Men's cells have to do it all the time. And the finding that XO women's cells do not contain a Barr body explains how they pull off this feat. Women with Turner's syndrome simply do not undergo X inactivation. Their *Xist* counts their X chromosomes, finds that there is only one, and then does nothing, just as it would in a man. No X inactivation, no Barr body, no X "underdosing," and no division of the body into two unequal populations. XO women are, in short, not mosaics like other women.

So you can see that it is the flexible nature of X inactivation that allows XO women to survive. And XO women are the proof that X inactivation does not occur because cells are part of a female-looking body, but because there are two X chromosomes. They also run contrary to the idea that X-borne diseases are "sex-linked." Although they look female, Turner's syndrome women are just as likely to get diseases like muscular dystrophy as men, and for exactly the same reason: they have only one X.

One thing scientists are not quite sure about is why women with Turner's syndrome look any different from XX women. If X inactivation is so thorough, then why should a woman with one X look any different from a woman with two? Scientists now think they know why this is. Perhaps X inactivation is not completely comprehensive—maybe not all the X-borne genes get switched off. You may remember that the X chromosome does share (and swap) a little region with a complementary area on the Y—the non-sexlike region. Any genes present in this region would be present in two copies in both XY and XX individuals, so maybe they never have to be inactivated at all. A problem would then only arise in XO women, because their cells each would have only one copy of these genes—and so it is thought likely that the characteristics of Turner's syndrome are caused by an "underdose" of the genes in this non-sexlike region.

Indeed, the effects of Turner's syndrome are now thought to be much more severe than previously believed. It has been estimated that XO embryos are formed much more commonly than the birth of XO babies suggests. Approximately one in fifty induced abortions are XO embryos, and as many as one-fifth of spontaneous miscarriage involves XO embryos. When statisticians got to work on these figures, they concluded that the XO girls who are born are just the tip of the Turner's iceberg—perhaps 97 percent of XO babies die in the womb. So being XO is usually fatal, and only the few mildest cases even make it as far as birth.

There are a few ways in which XO embryos could be created—a sperm without a sex chromosome entering a normal egg, a normal X-bearing sperm entering an X-less egg, or the embryo losing a sex chromosome soon after fertilization. Of all these, it is thought that the first possibility is the most common, but probably all can occur from time to time. Worryingly, this suggests that the whole process of fertilization is quite fragile, and it has been suggested that artificial fertility procedures such as ICSI (intracytoplasmic sperm injection) can disrupt it further, leading to an increased incidence of Turner's syndrome.

In fact it now appears that the creation of Turner's syndrome babies is not as simple as we once thought. On further examination, it transpires that many children with XO-like characteristics are not simple cases of the syndrome. Instead, some of them are actually a mixture of XO cells and XX cells, or a mixture of XO and XY cells. These children are mosaics, but of a slightly different kind to common-or-garden women. Probably, they are produced when an X or Y chromosome is accidentally lost in just some of the cells of a developing embryo. Rather neatly, their appearance seems to correspond fairly well to what one would expect—XO/XX mosaics appear intermediate in appearance between XO women and XX women, whereas XO/XY children can look very male. And the degree to which they differ from the "standard" XO appearance depends on the relative mixture of different cell types in their body.

And even in these Turner's mosaics, the process of X inactivation is faithfully working away. Whereas XO/XY boys are just simple mixtures of two cell types, with some XO cells and some XY cells, XO/XX girls are a mixture of no less than three types of cell. These girls contain XO cells, XX cells that have inactivated the paternally inherited X and XX cells that have inactivated the maternal X. No matter how complicated your genetic makeup, *Xist* still likes to keep your chromosomal house in order.

Another aspect of Turner's syndrome that has been reassessed in recent years is fertility. For a long time after its discovery, geneticists assumed that XO women were infertile—mainly because their ovaries do appear very small and inactive. Also, it was not clear how an XO precursor cell could embark on the complex chromosomal dance required to make normal X-bearing eggs. Yet many XO girls do in fact show signs of ovarian activity—they often go through some of the changes associated with puberty, for example. These changes are largely driven by female sex hormones, and these are only made to any great extent by developing egg follicles in the ovary, so there must be some functional eggs there. Careful examination of XO girls' ovaries has recently demonstrated that they do have active follicles containing apparently normal eggs. The ethical problem is that these eggs often disappear during the girls' teens, and so it is arguable whether these young girls should be encouraged to consent to surgical harvesting and storage of these eggs, as this is their only hope for producing children later in life.

The slightly haphazard approach to the number of X chromosomes in a cell works the other way as well: cells can easily cope with excess X. One of the commonest forms of sex-chromosome abnormality occurs in women who have three, four, or five X chromosomes in every cell in their body. Probably created by some failure of X chromosomes to separate during the chromosomal dance of the egg or sperm, X inactivation really comes to the aid of these women. *Xist* very thoughtfully counts up the X chromosomes in their cells, and inactivates all except one. As a result, XXX women, for example, have two Barr bodies and exactly one active X in every cell (and they are mixtures of three different cell types).

The net result is that these women are almost indistinguishable from XX women—and indeed many of them never realize that they are slightly unusual. Perhaps one in a thousand women are XXX, XXXX or XXXXX, but they can be hard to find. These "superwomen" do differ from XX women, but not so that you would notice. They are marginally less fertile than XX women and there is a contentious suggestion that they are slightly more likely to have learning difficulties. I think it is reasonable to suggest that these rather subtle effects may be due to some sort of "overdose" of that non-sexlike region of the X that does not get inactivated.

This playing fast and loose with the number of X chromosomes also extends to men, and perhaps one in five hundred are born with extra X chromosomes. The commonest permutation of this, first reported four years after Turner's syndrome in 1942, is XXY or Klinefelter's syndrome, but boys can also be born XXXY, XXXXY, XXYY, and XXXYY. Although Klinefelter's seems to have more effect on the male body than XXX-ness does on the female, it can still be so mild that many men do not realize they have it. Although it can sometimes result in small testicles, breast growth, unusually long legs, less body and facial hair, and lower intelligence, Klinefelter's can often be very subtle. Indeed, only now are many cases being diagnosed as more men are being investigated by infertility clinics—XXY men usually produce no sperm at all.

The relatively mild symptoms of Klinefelter's syndrome are probably caused by our constant friend, X inactivation. The *Xist* in an XXY man does not seem to care that there is a Y around—it simply gets on with its job of counting the X chromosomes and switching off all except one. As a result, Klinefelter's men have a Barr body in every cell in their body, and are mosaic mixtures of two dissimilar cell types—just like women (XXY tom cats can be calico). As we have seen, they are more prone to the civil war of autoimmune disease, and they can also asymptomatically "carry" sex-linked diseases such as muscular dystrophy. Most remarkably of all, their X inactivation is "reversed" in the testicle, just as it is in a woman's eggs, in anticipation of the production of sperm that unfortunately will never appear. Klinefelter's men are, in short, perhaps the greatest accidental wonder of X chromosome engineering. And to confuse matters further, some of them can even be XXY/XY mosaics as well.

One thing that is clear is that sexuality is far more complex than we ever realized. Geneticists' understanding of sexual differentiation has much advanced since Henking pondered his wallflower chromosome. Human beings are not simply male or female. In fact, all those ancient myths of male-female duality have turned out to be, well, mythical. "Intersexuality" is a fact of life, and not a particularly rare one either. Even when babies acquire so-called "normal" XX or XY chromosomal complements, later stages of the sex-determination process can go wrong, and all in all these bring the total up to about 1 percent of humanity who have a sexuality that differs from XX female or XY male. The human sexes are not two opposites, or even two equivalents—the human sexes are many and varied, and while two predominate, the others form a continuous spectrum of gender that stretches between those two, and beyond them.

Daddy's Girl

As you reach the end of this book it is worth reconsidering the importance of the X chromosome. It was discovered by chance, because although it looks pretty much like any other chromosome, it seems somehow different, distant from the others. Unlike the brash but belittled male-making Y, it has great depth and complexity that make it a far more interesting controller of our destinies. It brings death and destruction to lone-X men and complexity and confusion to dual-X women.

Obviously, although it seems to be one of the things that makes the sexes different, bear in mind that all this chromosomal wizardry only evolved so that males and females could get together and make some babies. Whatever difference and even conflict there is, sex is essentially a cooperative venture. And so to finish, I will briefly mention a way in which the X chromosome brings men and women together to achieve their goals.

All through this chapter I have described how women deal with the double bounty of X chromosomes that they inherit using X inactivation. Of course, the double life they lead does cause some problems, but in general the system works. Yet it now appears that inactivation is not always the best method of controlling the way that the X chromosome is used, and the evidence for this comes from a fascinating discovery in girls with Turner's syndrome.

It has been known for some time that some XO girls can be rather socially disruptive. Not to an alarming degree, but perhaps to the extent that they are as antisocial as many little boys. This may not strike you as surprising—after all, it could be just another part of the pattern of Turner's syndrome. And yet, there is an important fact at the center of this behavioral phenomenon. The XO girls who are disruptive are the ones that carry their mother's X. XO girls with the paternal X are socially fairly normal.

Why should the parental origin of their X chromosome make any difference? After much debate, the evidence has led geneticists to a rather strange conclusion—that there is a gene on the X chromosome that confers extra social skills on girls, over and above those of boys.

This is where things start to become weird: it seems that girls can only use the copy of the gene that they inherit from their father. So maternal-XO girls are disruptive and paternal-XO are not. And girls' behavior is controlled by a gene inherited from their father, but which their father could not use. And the gene cannot be X-inactivated into silence, as the paternal copy must always be available for use. And so on.

You may have realized by now that this female-sociability gene cannot be subject to the same process of X inactivation as most other X-borne genes. True X inactivation is fairly random, whereas the paternal copy of this gene is used preferentially. In fact, this gene is one of several that seem to escape X inactivation in some way—because although the inactivation of the X chromosome is fairly comprehensive in mice, the same is not true of women. Regions of the human X escape inactivation, and because of this the genes in these regions can regulate their use in different ways. And the sociability gene is one of these.

This exquisite theory raises a very big issue. Why should mothers and fathers conspire to stop inactivation of this well-behaved-girl gene in this way? And why should they then "agree" that it is the father's gene that gets used? Actually, this process of using only the gene from one parent has been reported before, but usually in genes carried on non-sex chromosomes, and often in rather different circumstances. It now appears that certain genes in eggs and sperm are tagged with a kind of marker that will allow the offspring to tell who they got those genes from. A kind of parent-specific chemical "imprint" is made on the genes, and so this whole process is called "genetic imprinting."

Genetic imprinting is not a widespread phenomenon in humans—offspring are usually blissfully unaware whence each gene comes—but when it does occur, it is usually a sign of tension between parents. In fact, imprinted genes are almost always used before birth, as part of a mother-father conflict over the development of the baby. To cut a long story short, fathers seem to want their babies to be simply as big as possible, so paternally inherited genes switch on to make the embryo grow as fast as possible. However, mothers have other priorities. Of course they want a healthy baby, but they do not want it to be so big that it exhausts or damages them so much that their future childbearing potential is compromised. So maternally inherited genes usually act to moderate the growth of the unborn child. This is why maternal and paternal genes behave differently from each other in growing babies: the parents slog it out for control of the child's growth.

Yet the imprinting of the girly behavior gene on the X chromosome is a very different affair, and far more amicable. For a start, it is very unusual for imprinting to affect people after birth, but this unusualness probably reflects the likelihood that, in this case, imprinting is a sign of parental cooperation, not conflict. Presumably, it is in both parents' interests that their sons behave like sons, and daughters behave like daughters—Daddy's little girl, even. And because this is an X-borne gene that is only supposed to act in daughters, then there is a perfect way to make sure that girls get exactly one active copy of it. Only copies of the gene inherited from fathers will function, since all girls have one father.

So little girls are sugar and spice and all things nice because their fathers made them that way. Maybe the gene that does this was used by their paternal grandmother, but it lay dormant in their father. The X chromosome has ingrained a delicious asymmetry between men and women, but a benign one in this case. It may be the Y chromosome that makes the obvious difference between men and women, but it is the X that makes them complementary rather than opposite. It is the X that eventually reunites them.

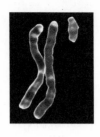

EPILOGUE:
THE CHOSEN ONE

Throughout this book I have described how the X chromosome is what sets the sexes apart. Well, I grudgingly accept that the vestigial shriveled Y may play a minor role in actually making embryos turn into boys, but I hope that by now I have convinced you that the recurring theme of difference between men and women is the predominance of the X chromosome. Although the Y writes the book, the X speaks the lines, as it were.

All in all, it is rather shocking how much of what we are is decided by the lottery of which sperm first arrives at the egg. And indeed, almost as soon as the mechanism of human sex determination was first elucidated, scientists started to wonder if they could cheat at this ultimate biological game of chance. If the sex chromosome that is carried in that sperm is the gobbet of life that divides us into male and female, sire and dam, single-X defective and double-X mosaic, then can we control this division? Can we use what we know to control the sex of our own children?

For decades scientists have been trying to find a practical way to control the sex of offspring. Some have claimed simple scientific inquisitiveness, others an agricultural motive (why produce unwanted dairy bull calves?), and a few others have openly stated their wish to allow human couples to control the sex of their children. Whatever the reasons given, few branches of biology have been more replete with blind alleys and dubious claims.

For several decades it has been clear that X- and Y-bearing sperm have enticingly different characteristics. One of the first differences reported was that X and Y sperm have different electrical charges, and it is even claimed than the egg is more receptive to one or other sperm type as its own charge "oscillates" over time. Whatever the validity of this idea, it certainly appears that separating sperm between electrodes does not leave them in a very healthy state.

For some time it has also been alleged that sperm may swim differently, depending on the identity of their sex chromosome passenger. Y sperm may be more likely to swim upward in a test tube, or X sperm may chart a straighter course in evenly flowing fluid. Y may penetrate certain gels better, and X may be purified by passing sperm though albumin. Or maybe these effects are caused by differences in shape or size. You might expect X sperm to be slightly larger than Y sperm, as they carry a heavier sex chromosome, but I cannot explain why the heads of X sperm are more circular than those of Y sperm.

It is quite surprising, however, that there are any discernible differences at all between the appearance or behavior of boy-genic and girl-genic sperm. Scientists had always considered sperm to be simple, efficient transporters of carefully packed genes. Indeed, chromosomes are more tightly packaged in sperm than in any other cell in the body—anything to make the sperm just that little bit more compact, it seems. Biologists long assumed that this systematic compression of sperm chromosomes meant that the chromosomes could not actually be used for anything in transit to the egg. The idea that sperm cannot use their own genes was lent further credence by the fact that sperm seem to need considerable help from other cells in the testicle to coordinate much of their own development—it is almost as if they are genetically inadequate in some way. But if sperm do not use their own genes, then why should X-bearing sperm behave any differently from Y-bearing sperm? An X sperm may contain many more genes than a Y sperm, but if they cannot be used, then that should make no difference. This remains a major paradox of sexual biology, as the two types of sperm clearly do differ.

One of the most frequently claimed ways to control the sex of babies is by paying a rather unhealthily precise attention to the timing of sexual intercourse. Time and again, the suggestion has been made that "early sex"—early, that is, in relation to ovulation—tends to lead to the conception of girls, and "late sex" produces more boys. How this might relate to the tradition that brides, and not grooms, can be late for their weddings, I will leave for you to consider. Of course, this theory sounds suspiciously like an old wives' tale, but like many such tales it now appears to contain a grain of truth. Studies in sheep have shown that timing of artificial insemination can have definite effects on the sex ratio of lambs conceived, and this effect has even been narrowed to the timing of sperm entry relative to certain developmental preparations going on in the egg.

If the timing of fertilization really does affect the sex of human babies, there could be several reasons why. It may be that one type of sperm swims faster, or is able to hang around longer, waiting for its eggy bride to appear. Perhaps the egg itself has some say in the matter, preferentially accepting the entry of one type of sperm when it is young and fresh, and the other type of sperm when it is old and jaded. Or possibly the time of the female cycle at which fertilization occurs affects the chances of survival of female and male embryos differently. Whatever the cause, there is one overriding fact on which most reproductive biologists are agreed: the effect of the timing of intercourse on a baby's sex is slight. If it happens at all, the chances of getting one sex or the other are only affected by a few percent at most, and even that only occurs if you know exactly when a woman has ovulated. However, the phenomenon may have very real relevance as more and more people receive fertility treatment, because one of the inevitable consequences of these treatments is that they disturb the natural chronology of the female cycle. For example, treatments to artificially induce ovulation have been claimed to result in more girl babies, whereas the use of artificial insemination may increase the chances of having a boy.

So, to cut a long story short, this wizardry in timing intercourse does not really lend itself to commercial exploitation—it is simply not reliable enough—and it has instead been co-opted by control-freak future parents. Anyway, when all is said and done, half of them will get the baby they want, and the other half can presumably blame their partner for not sticking to the agreed protocol of coital synchronization.

It now appears that you may be able to control your children's sex by timing things other than the actual act of copulation, although to do so you might have to make some difficult life decisions. This is because there is now evidence that the relative ages of parents can influence the sex of their children. A research group working at the University of Liverpool were fascinated by two particular pieces of demographic trivia. First, more boys are born during and immediately after wars—quite conveniently, since most people killed in wars are men. And second, in England and Wales it is a recorded fact that the age difference between husbands and wives became greater during these periods, with women tending to marry older men. These scientists could not help wondering whether these two facts were linked, and indeed, careful statistical scrutiny of birth records confirmed that having an older sexual partner makes a woman more likely to have boys.

This remarkable result, although subtle, may tell us a surprising amount about what makes us tick. Why ever should the age difference of a couple affect the sex of their children? One of the most challenging explanations for this phenomenon relates to the social status of men. There are several known examples in nature where animals produce different numbers of male and female offspring depending on their social status. One example is the red deer, in which it has been suggested that dominant females tend to produce male calves. The reason for this may be that in a polygamous species like red deer, a dominant stag son can pass his genes on to many many grandchildren.

In human society, female members of social elites often pair off with men considerably older than themselves—and have more sons than daughters. A fun anecdotal example of this is American presidents, who have, on average, sired considerably more sons. Could it pay for successful men to have sons, who will in turn be successful? Do successful men make their Y sperm better swimmers, or do women actively promote the survival of embryos of a certain sex because they think they have a good "catch"? And does this imply that human pair-bonding is largely a matter of female choice? I love all these hypothetical questions, because they allow hours of politically incorrect discussion, with little fear of any proof appearing to spoil the argument. Yet, as in the case of the timing of sex, the effect is subtle, and unlikely to result in the wholesale divorce and remarriage of large sections of the population just so they can have children of the desired sex.

As many readers will be aware, there is now a fairly effective way for parents to have a say in the sex of their children, and this is based on our heroine, the X chromosome. Over the last few years, fertility clinics have begun to advertise claims that they can dramatically skew the probability of couples having babies of a particular sex. And their claims are, on the whole, substantiated.

Their methods involve the separation of sperm into Y-rich and X-rich populations, which can then be used for fertilization of a waiting egg. Because the X chromosome is so much bigger than the Y, an X sperm contains considerably more DNA than a Y sperm, and it has been suggested for some time that this difference could be used to separate the two. In the early 1980s, reports appeared that in voles and chinchillas (mammals with especially large size discrepancies between X and Y), the two populations of sperm could be separated by a procedure called "flow cytometry." This involves firing sperm past a laser, measuring the light passing through and reflecting off them. If necessary the sperm can also be stained with a fluorescent dye that binds to DNA. The whole machine is hooked up to a computer, which rapidly works out the DNA content of each sperm as it flies by, and then diverts it into one of two containers—one for X and one for Y.

This technique may sound too good to be true, but it really does work. At first, the sperm were completely destroyed by the procedure, but as methods improved, they survived in a fit state to be injected into eggs, until today they can survive the procedure still able to make their own way into an egg. Also, the technique has been honed sufficiently that it works in species in which the X/Y size discrepancy is less than it is in voles and chinchillas, including humans. Currently, the process can produce a tube containing 90 percent human X sperm, or 70 percent Y sperm—impressive deviations from the normal fifty-fifty ratio.

This process is now commercially available in the United States, and has been producing babies for a few years now. Yet this ability to choose a child's sex falls squarely into the grey area between what society considers right and wrong. I would be the first to argue that people often simply fear the unfamiliar, but I do think that the facility the X chromosome has given us to control a baby's sex has also uncovered a very deep debate about what is, and what is not acceptable.

Perhaps I should make clear at this point that I have little time for arguments that sex selection is "playing God" or "simply against nature" and hence inherently wrong. The whole of medicine is equally "playing God" or "simply against nature," and yet few people want to ban medicine. Curing childhood leukemia is a profoundly unnatural thing to do, but I would not want to stop doctors doing it. To me, much so-called medical ethics is simply a reflection of the delay between the development of new procedures and the public's acceptance of them as familiar and commonplace.

Yet having ruled out the fear of novelty and the inconsistently applied distrust of the "unnatural," there are still two potentially valid reasons why sex selection of human children should be approached with caution. The first is that sex selection may somehow pervert the normal processes of society. Could an unhealthy obsession with having babies of a particular sex taint our perceptions of the value of human life? In facilities where sex selection procedures take place there appears to be little, if any, systematic bias toward couples wanting any particular sex. The emphasis is always claimed to be—and I have no reason to disbelieve these practitioners—on "family balancing." Essentially this means helping couples with male children conceive a girl, and vice versa. Indeed, clinics can have a policy of not helping couples conceive their first child, or conceive a child of the same sex as previous children. This approach is "gender neutral" as long as couples with girls and couples with boys both approach the clinics in equal numbers.

But is there a risk that this neutral approach will not be applied everywhere that sex selection technology becomes available? In many parts of the world, an altogether more sinister form of sex selection is underway. In Tamil Nadu in India, for example, in a society in which boys are especially prized, it is thought that one-third of all deaths of newborn babies are caused by selective neglect of girls. In one study, girls born to women who already had sons were reported to be 1.8 times more likely to die than boys, and girls born to women with no sons were 15.4 times more likely to die. This is neglect, but in some countries it is replaced by deliberate female infanticide—a scourge made worse by the state-imposed one-child policy of China, for example. It is unlikely that a sex-selection clinic will be set up in these areas in the next few months, but who knows what might happen in twenty years?

I would argue that pre-conception sex selection might actually be a good idea in societies in which one sex is especially prized—better to throw a tube of X-rich sperm in the trash than stand aside while a baby girl expires. The resulting uneven sex ratio of the population would of course have dramatic social ramifications, but maybe they would not be entirely negative. As the number of women declined, parents would soon realize that their best chance to produce grand-children might be to have daughters. In the societal sex market, women would become a scarce commodity, and their perceived value would increase accordingly. Eventually, you might even expect that an equilibrium would be reached at which the absolute need of a popu-lation for women would balance the society's ingrained concept of their "value." This is not really social engineering, as the society ends up finding its own solution regardless of the wishes of the sex-selection clinician. Much current suffering and death might be avoided, and there is less scope for concerted Frankensteinian abuse of the technology than one might expect.

This leaves the second, and to me the most worrying, aspect of sex selection—the effect it might have on the individuals in the families concerned. This is an issue that strikes at the heart of parents' deci-sions to try and conceive a child of a particular sex, even if it is for the apparently innocuous reason of "family balancing." I must stress that, in the case of couples who wish to avoid children of a particular sex for sound medical reasons—such as the fact that the woman carries an X-borne sex-linked disease—these concerns largely evaporate, but

for most other sex-selecting couples they are very real. And they relate
to the children, both planned and previous.

All technical procedures fail sometimes, and this is especially true
when technology strays into the messy terrain of biology. We have
already seen that sex selection is currently producing 75–90 percent
pregnancies of the hoped-for sex. This is quite a good success rate,
and no doubt it will improve in the future, but it means that 10–25
out of every 100 such pregnancies will produce children of the
"wrong" sex. How will families deal with these "failure" children? No
doubt they will often be loved and cared for as much as their same-sex
predecessor siblings, but how often will children grow up thinking
themselves unwanted? Even the hint of disappointment in their parents could be damaging to a child, and at worst it might sow a seed of
uneasiness that could corrode a family's happiness, even before the
"failure" child is old enough to sense that he or she is not quite wel-
come. Of course couples will be offered all sorts of counseling when
planning sex selection, but this cannot overcome the fact that some
will get what they want and others will not.

And even if couples do get what they want—a girl, for example,
after a long line of boys—another problem arises. How are these older
siblings to feel about the value of their own sex when their parents
were so eager to conceive their "chosen one," a child of the opposite
sex? And how will this uncertainty be compounded in the youngest of
those naturally conceived children—the last of the interminable row
of same-sex offspring that eventually pushed its parents over the line
into the fraught procedure of sex selection? There are parents, there
are the children they already have, there are the children that they
want, and there are the children they may end up with if the procedure does not work. I have no doubt that in some circumstances, all
can be treated honorably, but this might take a lot more thought than
it may seem at first. I believe that sex selection should be allowed, but

with exquisite attention to everyone involved, be they present at the outset, or produced by the process.

This is not to say that we should neglect the rights and wrongs of the issue, but I think that there is a final very good reason why sex selection poses such a finely balanced, and possibly insoluble, dilemma. Many people, and indeed most medical establishments, support the idea that in many circumstances, "negative selection" of children is ethically acceptable. For example, in many developed countries parents are offered the opportunity to have their unborn child tested for conditions such as Down's syndrome and spina bifida, with pregnancy termination a likely outcome if the diagnosis is positive. In fewer years than most of us think, this principle of negative selection will undoubtedly be extended to the diagnosis of genetic diseases in very early embryos—and few could claim that not continuing the progress of a ball of cells with a genetic defect is worse than terminating a pregnancy at a much more advanced stage of development. The principle is the same: to varying degrees, many of us accept the disposal of compromised embryos.

However, far fewer are willing to accept the converse—that embryos and babies should be "positively selected" for genetic traits that we consider beneficial. Of course there are some straightforward genetic objections to this process, including the fact that scientists do not yet understand much about how genes interact. For example, it is entirely possible that a geneticist could mistakenly select an embryo with a gene that makes the child both beautiful and prone to some terrible disease. Yet even when our understanding increases and these problems are ironed out, we are still left with a simple, central objection to positive selection. Our whole society is unwittingly based on the fact that we inherit genes in a haphazard way, and it is the variety which this engenders that makes our lives so interesting and satisfying. People are different, and they are different largely because they are not planned. Do we really want to lose this chance element in our lives? Do we really want to stop playing that delicious game?

So negative selection seems acceptable to many, just as positive selection is unacceptable. This is why I think it will be so difficult for society to make a decision about sex selection. It is not the exclusion of an unwanted disease, nor is it the choosing of a universally desirable trait. To the outside world it is a neutral choice between two equally valid outcomes—boy and girl. But to the parents involved, is it a process of positively picking the child they want, or a negative avoidance of the child they do not? Or can it be both at the same time?

The X chromosome has given us the ability to do what many humans, for right or wrong, have wanted to do for centuries—to choose the sex of their children. Whether we think sex selection is acceptable, and whether or not we are making that assessment for the right reasons, the procedure is here, and it is growing more popular. All I can ask is that we all consider our motives should we decide to select our children's sex. I hope this book has shown you that men and women are not opposites, not antagonistic, but simply very different for all sorts of reasons. The nature of our species is that we exist in two predominant forms, two sexes who can revel in each other's strangeness. Yet we must always bear in mind how our lives can go horribly wrong when we ascribe different values rather than valued differences to girls and boys.

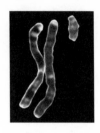

FURTHER READING

1. MAKING A DIFFERENCE

Of all the material in the Further Reading, this paper, first describing the discovery of the X chromosome, is probably the most difficult to find: H. Henking, *L. Zeit. Wiss. Zool.* 51 (1891).

Essence of Man, Essence of Woman

Aristotle's *Generation of Animals* is just about the most entertaining way to make a start in the world of reproductive biology—to read the thoughts of an intelligent man with little more but opinion and the naked eye to guide him. Follow this with William Harvey's equally marvelous book from the seventeenth century, *On the Generation of Animals,* and you will really see whence modern biology came.

And Here's the Reason Y

Here are two good descriptions of the history of genetics before and during the discovery of sex determination. The second is rather fanciful in places, but then again, there are not many contemporary records of Gregor Mendel's life: S. F. Gilbert, *Journal of the History of*

Biology (Fall) (1978): 307; R. M. Henig, *A Monk and Two Peas* (London: Weidenfeld and Nicolson, 2000). And if you want to claw your way through the original papers at the heart of our understanding of sex determination, then here they are; do be aware that, in common with most cutting-edge work, they are very much works in progress and contain some misinterpretations: C. E. McClung, *Biological Bulletin* 3 (1902): 43; also *Anatomischer Anzeiger* 20 (1901): 220; N. M. Stevens, *Journal of Experimental Zoology* 2 (1905): 371; E. B. Wilson, *Science* 22 (1905): 500.

The Great Chain of Being ... Sexy

Two papers in particular established the idea of the active production of the male embryo. The first is the discovery of the testicle as prime mover in the sexualization process, and the second is the discovery of *Sry*. See A. Jost, *Archives of Anatomy, Microscopy and Experimental Morphology* 36 (1947): 271; and A. H. Sinclair et al., *Nature* 346 (1990): 240. Filling in some of those enticing gaps in both the chain and the human life cycle are: B. Capel, *Annual Review of Physiology* 60 (1998): 497; C. M. Haqq, C.-Y. King, E. Ukiyama, S. Falsafi, T. N. Haqq, P. K. Donahoe, and M. A. Weiss, *Science* 266, 5190 (1994): 1494–1500; S. Nef and L. F. Parada, *Nature Genetics* 22 (1999): 295. Ursula Mittwoch has ploughed a completely different furrow in the field of sex determination. She claims that gonads become testicles because they are growing fast, and that ovaries are gonads that grew slowly. Although her theories are not widely accepted, they do help to point out some embarrassing gaps in the standard model of sex determination: U. Mittwoch, *Human Genetics* 89 (1992): 467; and *Journal of Experimental Zoology* 281 (1998): 466.

So Girls Are Just an Afterthought?

Well, are they? Nothing provokes the more philosophical of scientists and the more scientific of philosophers as this old chestnut. A great starting point is the scientifically accomplished book by Simone de Beauvoir, *The Second Sex* (New York: Knopf, 1949; 1953 English translation). And do not be put off by the fact that the following feminist critique of X and Y is in a scientific journal; it is an argumentative gem lent credibility by being written by a scientist who knows what she is talking about: J. A. Marshall Graves, *Biology of Reproduction* 63 (2000): 676.

Strong Woman

The idea that all may not be as it seems, and that the X chromosome may actually contain some female-making genes, is given full rein in S. S. Wachtel, *Cytogenetics and Cellular Genetics* 80 (1998): 222.

The Birds and the Bees

There is literature aplenty about the weird and wonderful world of sex determination in different animals. A brief survey will show that I have only scratched the surface in this book—if you can imagine a way of choosing your sex, then some animal, somewhere, will be using it. These three articles are about hermaphrodite fish, birds as backwards mammals, and mole voles, respectively: E. Clarke, *Science* 129 (1959): 215; H. Ellegren, *Trends in Ecology and Evolution* 15 (2000): 188; and Just et al. *Nature Genetics* 11 (1995): 117. The next three are more general, the first looking at the phenomenon of temperature-dependent sex determination and asking "why?":

R. Shine, *Trends in Ecology and Evolution* 14 (1999): 186; *Journal of Experimental Zoology* 281 (1998): 281ff; and J. H. Werren, *Annual Review of Ecology and Systematics* 29 (1998): 233.

Drifting Apart—the Sad Divorce of X and Y

These articles really represent the *Kramer versus Kramer* of the chromosomal world (why did the X and Y grow apart so?): M. L. Delbridge and J. A. Marshall Graves, *Reviews of Reproduction* 4 (1999): 101; K. Jegalian and D. C. Page, *Nature* 394 (1998): 776. And can they still interact? Some say yes and some say no: J. Weissenbach et al., *Development* 101 (1987): 67; T. Wiltshire, C. Park, and M. A. Mandel, *Molecular Reproduction and Development* 49 (1998): 70. If you want to go further, then you can read about the strange ways that genes waft around when they're held on the Y chromosome in M. J. Bamshad, *Nature* 395 (1998): 651. Or, for information about the quest to find the universal sex-determining gene: L. Meadows, and S. Russell, *Trends in Genetics* 17 (2001): 18.

INTERLUDE: WHAT IS IT, EXACTLY?

If you want to read about the discovery of the structure of DNA, and how it showed the way we inherit our genes, then the best source is the one written by someone present at the time: J. D. Watson, *The Double Helix* (New York: New American Library/Dutton, 1968). And Watson and Crick's original paper describing their interpretation of the structure of DNA from Franklin and Wilkins' samples is a remarkably brief and endearingly understated commencement to a scientific revolution: J. D. Watson and F. H. C. Crick, *Nature* 171 (1953): 737.

2. THE DUKE OF KENT'S TESTICLES

For more on the story of hemophilia, try G. I. C. Ingram, *Journal of Clinical Pathology* 29 (1976): 269; and for those remarkable ancient Jewish "references" to the disease, see F. Rosner, *Annals of Internal Medicine* 70 (1969): 833.

Genes that Jump Generations

How can a girl get hemophilia? See S. Windsor, A. Lyng, S. A. Taylor, B. M. Ewenstein, E. J. Neufeld, and D. Lillicrap, *British Journal of Haematology* 90 (1995): 906. Was Victoria a bastard? See D. M. Potts and W. T. W. Potts, *Queen Victoria's Gene* (Stroud, U.K.: A. Sutton Publishers, 1995). Two more general articles on the evolution of X-linked diseases are: K. Lange and R. Z. Fan, *Theoretical Population Biology* 51 (1997): 118; and C. Oudet, H. von Koskull, A. M. Hordstrom, M. Peippo, and J. L. Mandel, *European Journal of Human Genetics* 1 (1993): 181.

The Vulnerable Giant

The literature on muscular dystrophy is understandably huge. Here are several articles on the genetic basis of the disease, including some rare variants: A. H. Ahn and L. M. Kunkel, *Nature Genetics* 3 (1993): 283; S. M. Gospe et al., *Neurology* 39 (1989): 1277; E. P. Hoffman, R. H. Brown, and L. M. Kunkel, *Cell* 51 (1987): 919; Y. Sunada, and K. P. Campbell, *Current Opinion in Neurology* 8 (1995): 379; and H. Zellweger and E. Niedermeyer, *Annals of Paediatrics* 205 (1965): 25. The identical twin experiment that hunted for a treatment for muscular dystrophy is: M. Zatz, R. T. B. Betti, and O. Frota-Pessoa, *American Journal of Medical Genetics* 24 (1986): 549.

Men: Closer to the Apes?

Much of the most interesting research into color blindness compares our own vision with that of our primate relatives. The following are about howler monkeys and squirrel monkeys, respectively: K. S. Dulai, M. von Dornum, J. D. Mollon, and D. M. Hunt, *Genome Research* 9 (1999): 629; G. H. Jacobs and J. Neitz, *Proc. Nat. Acad. Sci. USA* 84 (1987): 2545. And for a discussion of why three-color vision may be so desirable in primates, see N. J. Dominy and P. W. Lucas, *Nature* 410 (2001): 363; P. R. Martin, B. B. Lee, A. J. R. White, S. G. Solomon, and L. Rüttiger, *Nature* 410 (2001): 933; and Y. Tan and W-H. Li, *Nature* 402 (1999): 36. On the hunt for the number of functional cone pigment genes, see S. S. Deeb, A. L. Jorgensen, L. Battisti, L. Iwasaki, and A. G. Motulsky, *Proceedings of the National Academy of Sciences USA* 91 (1994): 7262; M. Neitz and J. Neitz, *Science* 267 (1995): 1013; J. Winderickx, L. Battisti, A. G. Motulsky, and S. S. Deeb, *Proceedings of the National Academy of Sciences USA* 89 (1992): 9710. The strange story of the Pingelap color-blind is told in O. Sacks, *Island of the Colorblind* (New York: Vintage Books, 1998). Are color-blind men better at penetrating camouflage, and did this make color blindness more common in human populations? See M. J. Morgan, A. Adam, and J. D. Mollon, *Proceedings of the Royal Society of London B* 248 (1992) 291; and R. H. Post, *Eugenics Quarterly* 9 (1962): 131.

INTERLUDE: HOW SEXY IS X?

The original prediction that sex-related genes should congregate on the X chromosome was made in W. R. Rice, *Evolution* 38 (1984): 735, and seems to have been borne out by subsequent studies: K. Reinhold, *Behavioural Ecology and Sociobiology* 44 (1998): 1; G. M. Saifi and H. S. Chandra, *Proceedings of the Royal Society of London Series B* 266 (1999): 203; P. J. Wang, J. R. McCarrey, F. Yang, and D. C. Page, *Nature Genetics* 27 (2001): 422. But are genes sexy before they are attracted to the X, or do they become so afterward? See D. Charlesworth and B. Charlesworth, *Genetic Research* 35 (1980): 2–5. And what about genes on the Y chromosome? See B. T. Lahn and D. C. Page, *Science* 278 (1997): 675.

The scientific study of the putative genetic cause of homosexuality has been as tortuous and controversial as you might expect. For example: some maintain there is a "gene for homosexuality" on the Y chromosome; see R. Blanchard and P. Klassen, *Journal of Theoretical Biology* 185 (1997): 373. Or rather some believe it is on the X chromosome: D. H. Hamer, S. Hu, V. L. Magnuson, N. Hu, and A. M. L. Pattatucci, *Science* 261 (1993): 321. Or perhaps it is on neither: G. Rice, C. Anderson, N. Risch, and G. Ebers, *Science* 284 (1999): 665.

3. THE DOUBLE LIFE OF WOMEN

A Passable Mosaic

Three papers lie at the root of our understanding of X-inactivation: the discovery of the Barr body—M. L. Barr and E. G. Bertram, *Nature* 163 (1949): 676; the claim that it is, in fact, an X chromosome—S. Ohno and T. S. Hauschka, *Cancer Research* 20 (1960): 541; and the wonderfully succinct combination of this information with some observations of mottled mice—M. F. Lyon, *Nature* 190 (1961): 372.

Getting the Dose Right

There are two excellent general reviews of the how and why of X-inactivation: B. M. Cattanach and C. V. Beechey, *Development supplement* (1990): 63; and M. F. Lyon, *Biological Review* 47 (1972): 1. Recently, research in this field has concentrated on the discovery and characterization of *Xist,* the gene responsible for it all: C. J. Brown et al. *Nature* 349 (1991): 38; C. J. Brown et al. *Cell* 71 (1992): 55525; and G. D. Penny, G. F. Kay, S. A. Sheardown, S. Rastan, and N. Brockdorff, *Nature* 379 (1996): 131.

Exceptions that Prove the Rule

Some important deviations from the placental mammal system of random X-inactivation are our own fetal membranes—see T. Tada, M. Tada, and N. Takagi, *Development* 119 (1993): 813; marsupials—see D. W. Cooper, *Australian Journal of Biological Science* 29 (1977): 345; and eggs—see G. F. Kay, S. C. Barton, M. A. Surani, and S. X. Rastan, *Cell* 77 (1994): 639; and K. Mroz, L. Carrel, and P. A. Hunt, *Developmental Biology* 207 (1999): 229. And, as with everything in reproduction, there are almost as many ways of adjusting your X-dose as there are kinds of animal: see V. H. Meller, *Trends in Cell Biology* 10 (2000): 54.

Some You Win, Some You Lose

Dissimilarity between "identical" twin girls happens all the time; see J. Burn et al., *Journal of Medical Genetics* 23 (1986): 494. It can even lead to one twin having a sex-linked disease and one not; see M. R. Gomez, A. G. Engel, G. Dewald, and H. A. Peterson, *Neurology* 27 (1977): 537; C. S. Richards et al., *American Journal of Medical Genetics* 46 (1990): 672; and J. R. Lupski, C. A. Garcia, H. Y. Zoghbi, E. P. Hoffman, and R. G. Fenwick, *American Journal of Medical Genetics* 40 (1991): 354. Perhaps this problem can be overcome in some cases by transplantation of cells from the other twin? See J. P. Tremblay et al. *Neuromuscular Disorders* 3 (1993): 583. And two papers that disagree as to whether X-inactivation is the cause of identical twinning are: J. Goodship, J. Carter, and J. Burn, *American Journal of Medical Genetics* 61 (1996): 205; and G. A. Machin, *American Journal of Medical Genetics* 61 (1996): 216.

The Historical Origins of the Civil War

Why are women more prone to autoimmune diseases? Is it something to do with their fundamentally mixed bodies? See S. Chitnis et al., *Arthritis Research* 2 (2000): 399; and J. J. Stewart, *Immunology Today* 19 (1998): 352.

There Is Always Another Way

On the topic of maleness, femaleness, and all the other possibilities, see D. R. J. Bainbridge, *Making Babies* (Cambridge, Mass.: Harvard University Press, 2001); A. D. Dreger, *Hermaphrodites and the Medical Invention of Sex* (Cambridge, Mass.: Harvard University Press, 1998); and H. F. Klinefelter, E. C. Reifenstein, and F. Albright, *Journal of Clinical Endocrinology and Metabolism* 2(1942): 615. Do most Turner's syndrome babies die before they are born? See F. Hecht and J. P. Macfarlane, *Lancet* 2 (1969): 1197.

Daddy's Girl

I recommend two papers on the unequal use of genes because of the parent from whom they were inherited. Is genetic imprinting for parental conflict? See T. Moore and D. Haig, *Trends in Genetics* 7 (1991): 45. Or is it for parental cooperation? See Y. Iwasa, *Current Topics in Developmental Biology* 40 (1998): 255.

EPILOGUE: THE CHOSEN ONE

Does parental age difference affect the sex of babies? See J. T. Manning, R. H. Anderton, and M. Shutt, *Nature* 389 (1994): 344. Separating sperm certainly affects the sex of babies. See D. L. Garner et al., *Biology of Reproduction* 28 (1983): 312; and D. Pinkel, B. L. Gledhill, S. Lake, D. Stephenson, and M. A. van Dilla, *Science* 218 (1982): 904.

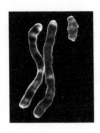

GLOSSARY

I have tried to avoid jargon terms throughout this book, but if you pursue some of the works mentioned in Further Reading, I hope that this glossary will help you.

Allele   See *gene*.

Amino acid   The chief components of a *protein*.

Androgens   A group of similar hormones, including testosterone, which drive the formation of the male body, as well as changes at puberty. They also drive sperm production.

Autosomes   *Chromosomes* that are not *sex chromosomes*, what I have called in this book non-sex chromosomes—in other words, most of them.

Barr body   A bundle of condensed *DNA* present in the *nucleus* of many cells in female mammals. It is the inactivated X *chromosome*. Also called *sex chromatin*.

Calico   A cat with a patchy coat, consisting of flecks of ginger mixed with flecks of another color, often black. The effect occurs in cats that are X *chromosome mosaics*—often XX females—because the ginger coat gene is carried on the X chromosome. See also *tortoiseshell*.

Carriers   Individuals who carry a single copy of a *gene* for an inherited disease, but who do not themselves show symptoms of that disease because they also have a normal copy of that gene.

Chimera   An animal that is made up of more than one genetically distinct population of cells, where these cells are derived from more that one fertilized egg. Compare this with *mosaic.*

Chromosomes   Paired structures in the cell nucleus that are made of densely packed *DNA.* As such they carry most of a cell's genetic material. Human cells usually have 46 chromosomes—44 *autosomes* and two sex chromosomes.

Cones   The *photoreceptors* in the eye that respond to bright light. Some animals have more than one set of cones, which allows them to discern colors.

Creatine kinase   A protein that spills into the blood from damaged muscle cells, as happens in muscular dystrophy.

Cryptorchid   Retention of at least one testicle in the abdomen. Next time you go to a flower show, just remember that "orchid" is Greek for "testicle."

Dichromat   An animal or person with two populations of *cones* that can respond to appreciably different wavelengths of light. See also *trichromat, monochromat.*

Dioecious   Animals that exist in two distinct sexes, like humans.

Diploid   Possessing two copies of each *gene,* like humans and female bees do—see *haploid.*

Dizygotic twins   Twins that form from separate fertilized eggs. Also called "fraternal" or "nonidentitcal" twins. Twins can also, more rarely, form from a single fertilized egg and these are called *monozygotic* or "identical" twins.

DNA   Deoxyribonucleic acid is a long, linear molecule that carries sections of code called genes, rather like a ticker tape. In animals and plants, the DNA is usually neatly packaged into chromosomes.

Dystrophin   The largest known mammalian *gene*. Carried on the X *chromosome*, it is damaged in the commonest forms of *muscular dystrophy*.

Exons   The regions of a *gene* that actually contain the code for the *protein* it encodes.

Gene   A gene is a piece of code written along the length of a *DNA* molecule. In animals, this DNA is packaged into *chromosomes*. This code is often translated by the machinery of the cell to make a *protein*. Humans inherit two copies of most genes—one from each parent. The two copies are called alleles. The particular position on the chromosome that carries the code for a particular gene is called a locus.

Gonad   General term for organs than make germ cells (either sperm or eggs), and in humans this usually means ovaries or testicles.

Gubernaculum   A strand of tissue attached to the testicle at one end and to the inside of the scrotum at the other. As a male fetus grows, the gubernaculum fails to keep pace and so becomes relatively shorter. As a result the testicles are drawn out of the abdomen into the scrotum.

Hemophilia   A disease in which the blood does not clot effectively, and sufferers often show prolonged bleeding after injury. Many forms of hemophilia are inherited genetic diseases. The two commonest forms of hemophilia are called hemophilia A and hemophilia B, and they result from damage to two genes on the X *chromosome:* factor VIII and factor IX, respectively.

Haploid   Possessing one copy of each *gene,* like male bees do—see *diploid.*

Hormone   A chemical made by one part of the body that enters the bloodstream (usually) and causes diverse effects in other, distant tissues.

Introns   The regions of a gene that separate the coding regions, the *exons.*

Locus   See *gene.*

Lyonization   Another term for the *mosaic* nature of *X inactivation.*

Monochromat   An animal or person with one population of *cones.* See also *dichromat, trichromat.*

Monozygotic twins   Twins that form from a single fertilized egg. Also colloquially, and erroneously, called "identical" twins. Twins can also, more commonly, form from different fertilized eggs and these are called *dizygotic* or "fraternal" twins.

Mosaic   An animal that is made up of more than one genetically distinct population of cells, where these cells are derived from a single fertilized egg. Compare this with *chimera.*

Mullerian inhibiting substance (MIS)   Also called anti-Mullerian hormone (AMH) or Mullerian inhibiting factor (MIF). A *hormone* made by the developing testicle that drives the formation of the testicle and the destruction of the female (Mullerian) reproductive ducts.

Muscular dystrophy   A disease involving progressive degeneration of the muscles of the body. The two commonest forms of muscular dystrophy occur because of damage to the *dystrophin gene* on the X *chromosome.* Duchenne is the commoner, more severe form, and Becker is the rarer, milder form. Other, rarer forms of muscular dystrophy are not a result of damage to genes on the X *chromosome.*

Nucleus   The central compartment of the cell, which contains most of its *DNA.*

Opsin   A *protein* that, together with a *retinal* molecule, forms the visual pigment of the eye.

Parthenogenesis   A form of asexual reproduction that involves female animals spontaneously conceiving daughters genetically identical to themselves. Parthenogenetic species are all-female.

Photoreceptors   The cells in the retina at the back of the eye that actually respond to light. Most people have four kinds: rods, which respond to dim light, and red, green, and blue cones, which respond to bright light.

Protein   A molecule constructed as a single chain of *amino acids,* built according to the code held in a *gene.*

Pseudoautosomal region   The region of the Y chromosome that occasionally exchanges fragments with the X chromosome. In doing this, it behaves more like part of a non-sex chromosome, or *autosome,* and so in this book I have called it the "non-sexlike region."

Retinal   A small molecule derived from vitamin A that, together with an *opsin* molecule, forms the visual pigment of the eye.

Rods   The *photoreceptors* in the eye that respond to dim light.

Sex chromatin   See *Barr body.*

Sex chromosomes   These *chromosomes* differ from all the others (called *autosomes*) because they are involved in some way in controlling the sex of the animal. In humans they are called X and Y, and in birds W and Z.

Sex-linked diseases   Diseases that are especially common in one sex (usually men) because they are caused by damage to a gene on one *sex chromosome* (usually the X). Examples include commoner forms of *hemophilia, muscular dystrophy,* and *color blindness.*

Sry   The *gene* now thought to trigger development of the testicle in mammals, and thus make embryos become male. Its name stands for "*s*ex determining *r*egion on the *Y* chromosome."

Tortoiseshell   Another term for *calico.*

Trichromat   An animal or person with three populations of *cones* that can respond to appreciably different wavelengths of light. See also *dichromat, monochromat.*

Turner's syndrome   A collection of characteristics often seen in babies born with only one sex *chromosome,* an X.

X inactivation   The process by which, in mammalian embryos containing more than one X *chromosome* (such as XX females), all X chromosomes except one are rendered inactive, destined to become *Barr bodies.*

X inactivation center (XIC)   The point on the X *chromosome* at which it starts to condense down into a *Barr body.*

Xist   The gene thought to cause the selection and switching off of the inactivated X *chromosome.* The name stands for "*i*nactivated *X* chromosome–specific *t*ranscript."

Zfy   The *gene* originally thought, erroneously, to be the trigger to maleness in mammals. Its name stands for "*z*inc *f*inger on the *Y* chromosome" A zinc finger is part of a *protein* that interacts with *DNA* molecules.

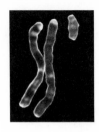

INDEX